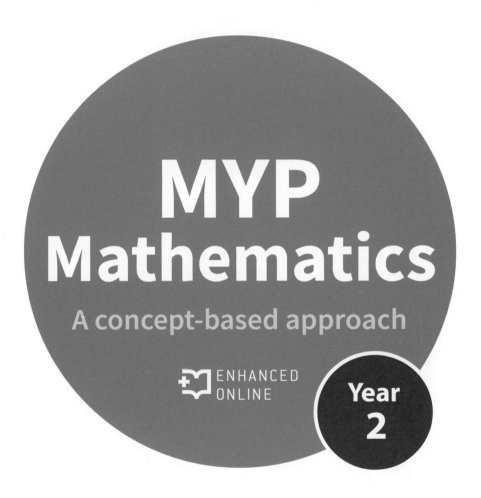

MYP Mathematics

A concept-based approach

ENHANCED ONLINE

Year 2

David Weber
Talei Kunkel
Alexandra Martinez
Rebecca Shultis

OXFORD

Great Clarendon Street, Oxford, OX2 6DP, United Kingdom

Oxford University Press is a department of the University of Oxford.
It furthers the University's objective of excellence in research, scholarship, and education by publishing worldwide. Oxford is a registered trade mark of Oxford University Press in the UK and in certain other countries

First published in 2018

British Library Cataloguing in Publication Data
Data available

ISBN 978-0-19-835616-5

11

Paper used in the production of this book is a natural, recyclable product made from wood grown in sustainable forests.
The manufacturing process conforms to the environmental regulations of the country of origin.

Printed in China by Golden Cup

Acknowledgements

Cover: Gil.K/Shutterstock
p5: Ollyy/Shutterstock; p6 (T): FatCamera/iStockphoto; p6 (B): GlobalStock/iStockphoto; p8: Lesley Vamos/Andrea Brown Literary Agency; p14: Gary Ramage/Newspix; p16: Sth/Shutterstock; p18: 3Dsculptor/Shutterstock; p19: Gregory Johnston/Shutterstock; p24: Igor Stramyk/Shutterstock; p25: Wojtek Chmielewski/Shutterstock; p28: Philip Bird LRPS CPAGB/Shutterstock; p30 (L): Elenabsl/Shutterstock; p30 (R): Baibaz/Shutterstock; p38: Joker1991/Shutterstock; p40: Natursports/Shutterstock; p41: Herbert Kratky/Shutterstock; p37: Koosen/Shutterstock; p45: Lano Lan/Shutterstock; p46 (TC): Stephane Bidouze/Shutterstock; p46 (TR): Monkey Business Images/Shutterstock; p46 (CC): Monkey Business Images/Shutterstock; p46 (CR): K. Miri Photography/Shutterstock; p46 (LC): StockLite/Shutterstock; p46 (LR): Ruslan Semichev/Shutterstock; p48 (CL): Brixel/Shutterstock; p48 (CC): Chrisbrignell/Shutterstock; p48 (CR): Mexrix/Shutterstock; p52: Boule/Shutterstock; p54: Appleboxcreative/Shutterstock; p57: Milosk50/Shutterstock; p59: Iakov Filimonov/Shutterstock; p60 (T): Farres/Shutterstock; p60 (B): Monkey Business Images/Shutterstock; p62: Iaroslav Neliubov/Shutterstock; p63: Grum L/Shutterstock; p53: Nicky Rhodes/Shutterstock; p55: Anneka/Shutterstock; p58: Gjermund Alsos/Shutterstock; p68: Everett Historical/Shutterstock; p78: Villorejo/Shutterstock; p87: Diane C Macdonald/Shutterstock; p91: L.Watcharapol/Shutterstock; p92 (TR): Andrey Armyagov/Shutterstock; p92 (T): Rangizzz/Shutterstock; p92 (TL): Iurii/Shutterstock; p95: Oleg Nekhaev/Shutterstock; p99: Rook76/Shutterstock; p103: Clavivs/Shutterstock; p104: A3D/Shutterstock; p105: Everett Historical/Shutterstock; p112: Yet to come; p121: Andy Hamilton; p124: Everett Historical/Shutterstock; p133: Rainer Lesniewski/Shutterstock; p136 (T): Vietnam Stock Images/Shutterstock; p136 (B): Repina Valeriya/Shutterstock; p137: Merlin74/Shutterstock; p141: Anneka/Shutterstock; p154: Happy monkey/Shutterstock; p160: Mega Pixel/Shutterstock; p166: Lagui/Shutterstock; p170: Creative-Material/Shutterstock; p172: Georgios Kollidas/Shutterstock; p187: Orin/Shutterstock; p188: Robert D Pinna/Shutterstock; p194 (T): Jeffrey M. Frank/Shutterstock; p195 (T): Sara Winter/Shutterstock; p194 (B): Globe Turner/Shutterstock; p195 (B): Phototreat/iStockphoto; p198: Nature Production/naturepl.com; p199 (T): Yann Arthus-Bertrand/Getty Images; p199 (B): Panuwatccn/Shutterstock; p200 (CL): Javier Trueba/MSF/Science Photo Library; p203: Pamela Au/Shutterstock; p204: Naeblys/Shutterstock; p208 (T): ArtOfPhotos/Shutterstock; p208 (B): Richard A McMillin/Shutterstock; p212 (T): Nora Yusuf/Shutterstock; p212 (B): Diyana Dimitrova/Shutterstock; p213: S-F/Shutterstock; p218 (T): Heritage Image Partnership Ltd / Alamy Stock Photo; p218 (B): Shootmybusiness/Shutterstock; p220 (TR): SeanPavonePhoto/Shutterstock; p219: Gennady Stetsenko/Shutterstock; p220 (BL): JPL/NASA; p220 (BR): NASA/JPL-Caltech/Space Science Institute; p222: Jbd30/Shutterstock; p227: Regien Paassen/Shutterstock; p228 (T): Pogonici/Shutterstock; p228 (B): Ph0neutria/Shutterstock; p229 (T): Milagli/Shutterstock; p229 (B): Steve Cukrov/Shutterstock; p230: Brocreative/Shutterstock; p231: Alexander Tolstykh/Shutterstock; p240 (TL): Vicki Vale/Shutterstock; p240 (TR): 7th Son Studio/Shutterstock; p240 (LT): Pretty Vectors/Shutterstock; p240 (RT): Kostenyukova Nataliya/Shutterstock; p240 (LB): SAYAM TRIRATTANAPAIBOON/Shutterstock; p240 (RB): Tusumaru/Shutterstock; p240 (BL): Maks Narodenko/Shutterstock; p240 (BR): Ungor/Shutterstock; p253: JPL/NASA; p254-255: Kak2s/Shutterstock; p259: Lledo/Shutterstock; p270: Mivr/Shutterstock; p275: Cathy Yeulet/123RF; p283: Rawpixel/Shutterstock; p289: Milles Studio/Shutterstock; p298: OlegD/Shutterstock.

Artwork by Thomson Digital.

Contents

Launch additional digital resources for this book

MYP Mathematics 2: a new type of math textbook...

This is not your average math textbook. Whereas most textbooks present information (kind of like a lecture on paper), which you then practice and apply, this text will help you develop into a mathematician. Following the MYP philosophy, you will perform investigations where you will discover and formulate mathematical rules, algorithms and procedures. In fact, you will generate the mathematical concepts yourself, before practicing and applying them. Much like an MYP classroom, this text is supposed to be an active resource, where you learn mathematics by doing mathematics. You will then reflect on your learning and discuss your thoughts with your peers, thereby allowing you to deepen your understanding. You are part of a new generation of math student, who not only understands how to do math, but what it means to be a mathematician.

Acknowledgements from the authors:

Talei Kunkel: *I would like to thank my husband, James, for his support and patience and my two children, Kathryn and Matthew, as they are my inspiration and motivation for writing this series of textbooks.*

Alexandra Martinez: *To my family and David Weber, for believing in me always.*

David Weber: *I would like to thank my father and Bobby, the two men who inspired me to never stop believing, as well as my students, who push me to always strive for more.*

How to use this book

MYP Mathematics 2 is designed around six global contexts. They are:
- identities and relationships
- orientation in space and time
- personal and cultural expression
- scientific and technical innovation
- globalization and sustainability
- fairness and development

Each unit in this book explores a single global context. However, a different global context could have been selected as a focus of study. What would the study of this content look like in a different context? The chapter opener gives you just a small taste of the endless possibilities that exist, encouraging you to explore and discover these options on your own.

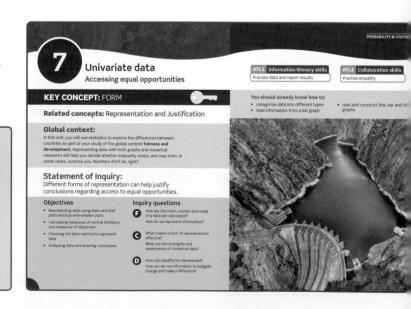

Topic opening page

Key concept for the unit.

The identified **Global context** is explored throughout the unit through the presentation of material, examples and practice problems.

Statement of Inquiry for the unit. You may wish to write your own.

The Approaches to Learning (**ATL**) skills developed throughout the unit.

You should already know how to: – these are skills that will be applied in the unit which you should already have an understanding of. You may want to practice these skills before you begin the unit.

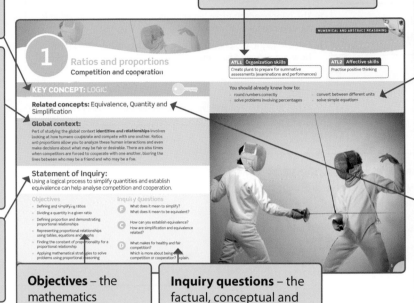

The **related concepts** explored in the unit. You may want to see if other related concepts could be added to this list.

Objectives – the mathematics covered in the unit.

Inquiry questions – the factual, conceptual and debateable questions explored in the unit.

Learning features

Investigations are inquiry-based activities for you to work on individually, in pairs or in small groups. It is here that you will discover the mathematical skills, procedures and concepts that are the focus of the unit of study.

Activities allow you to engage with mathematical content and ideas without necessarily discovering concepts. They allow you to practice or extend what you have learned, often in a very active way.

Activity 3 – Solving proportions: a different approach

Another way to solve proportions uses the techniques you learned to solve equations.

1 Solve the following equations.

$$\frac{w}{4} = 5 \qquad 10 = \frac{y}{3} \qquad \frac{x}{3} = \frac{21}{9} \qquad \frac{5}{8} = \frac{h}{16}$$

Investigation 3 – Solving proportions

Look at the following equivalent fractions/proportions.

$\frac{2}{3} = \frac{4}{6}$	$\frac{10}{20} = \frac{4}{8}$	$\frac{1}{6} = \frac{4}{24}$	$\frac{1}{4} = \frac{10}{40}$
$\frac{10}{15} = \frac{2}{3}$	$\frac{5}{6} = \frac{10}{12}$	$\frac{3}{4} = \frac{6}{8}$	$\frac{2}{5} = \frac{6}{15}$
$\frac{3}{8} = \frac{12}{32}$	$\frac{2}{7} = \frac{6}{21}$	$\frac{4}{9} = \frac{20}{45}$	$\frac{100}{250} = \frac{4}{10}$

criterion **B**

a Write down patterns you see that are the same for **all** pairs.

b Share the patterns you found with a peer. Which one(s) can be written as an equation?

c Generalize your patterns by writing one or more equations based on the proportion $\frac{a}{b} = \frac{c}{d}$.

d Add four more equivalent fractions/proportions to the ones here and verify your equations for these new proportions.

Hints are given to clarify instructions and ideas or to identify helpful information.

In Question **1**, parts **a** and **b** are order 3 perimeter magic triangles; part **c** is order 4.

Example 1

Q In physics, 'velocity' is a speed with a direction. It can be either positive (in this example, heading east) or negative (heading west). An airplane is traveling at −80 km/h down a runway.

a The pilot needs to triple her velocity in order to take off. What is her desired take-off velocity?

b Once in flight, the plane's cruising velocity will be −800 km/h. How many times faster than the runway speed (−80 km/h) is this?

A **a** $-80 \times 3 = -240$

To triple means to multiply by 3.

A positive number multiplied by a negative number results in a negative number.

Examples show a clear solution and explain the method.

Reflect and discuss boxes contain opportunities for small group or whole class reflection on the content being learned.

Reflect and discuss 7

- How does your rule from Investigation 6 relate to the rule you generalized for multiplying several integers? Explain.

- Is $-3^2 = (-3)^2$? Explain.

Practice questions are written using IB command terms. You can practice the skills learned and apply them to unfamiliar problems. *Answers for these questions are included at the back of the book.*

Practice 4

1 Find the following products.

  a -4×8 **b** $-3 \times (-7)$ **c** $10 \times (-6)$ **d** $-5 \times (-9)$ **e** -11×4

  f 8×6 **g** $-7 \times (-7)$ **h** -6×9 **i** $-2 \times (-1)$ **j** 7×5

  k -11×11 **l** $9 \times (-3)$ **m** $-6 \times (-6)$ **n** 4×12 **o** -12×4

  p $-5 \times (-2)$ **q** $2 \times (-12)$ **r** 10×7 **s** $-3 \times (-2)$ **t** -9×1

2 Match the expressions that have the same value.

-2×6 $10 \times (-2)$ $-3 \times (-4)$ 2×10

$-5 \times (-4)$ -3×4

-12×1 $-2 \times (-10)$

$-1 \times (-12)$ -20×1 -4×5

$-1 \times (-20)$ $-6 \times (-2)$ -4×3

▶ Continued on next page

Weblinks present opportunities to practice or consolidate what you are learning online or to learn more about a concept. While these are not mandatory activities, they are often a fun way to master skills and concepts.

For practice with estimating fractions and decimals graphically, go to brainpop.com and search for 'Battleship Numberline'.

Formative assessment 1

ATL1

On July 16, 1969, the Apollo 11 mission launched from Cape Kennedy, Florida, and, four days later, landed on the surface of the Moon. Neil Armstrong and Edwin 'Buzz' Aldrin spent just under 22 hours on the Moon's surface before rejoining Michael Collins in the command module and landing back on Earth on July 24, 1969. The mission was more than a race to the Moon; it also included lunar exploration and setting up monitoring equipment. Since then, many missions have gone to the Moon (or near it) to collect a wide range of data.

1 The force of gravity on the Moon is significantly less than it is on Earth. In order to approximate this force, multiply the mass of the object by −2. Remember, on Earth, the force of gravity is calculated by multiplying the mass by −10.

a Find the force of gravity on each of the following items, both on Earth and on the Moon. Show your working in your own copy of this table.

Equipment	Mass (kg)	Force of gravity on Earth (N)	Force of gravity on the Moon (N)
Lunar module (*Eagle*)	15103		
Neil Armstrong	75		
Command module	5557		
Samples of lunar rock	22		

b How many times larger is the force of gravity on Earth than the force of gravity on the Moon? Show your working.

2 On the surface of the Moon, temperatures and elevations vary widely. Since there is no ocean, elevations are given in terms of how far above and below the average radius of the Moon (1737.4 km) they are.

Forces are quantities that can be measured in a variety of units. The SI unit of force is the newton (N), named after Sir Isaac Newton, because of his study of motion.

Formative assessments help you to figure out how well you are learning content. These assessments explore the global context and are a great way to prepare for the summative assessment.

criterion **D**

Technology icon indicates where you can discover new ideas through examining a wider range of examples, or access complex ideas without having to do lots of painstaking work by hand. This icon shows where you could use Graphical Display Calculators (GDC), Dynamic Geometry Software (DGS) or Computer Algebra Systems (CAS).

ATL icons highlight opportunities to develop the ATL skills identified on the topic opening page.

ATL1

Each unit ends with

Unit summary recaps the key points, ideas and rules/formulas from the unit.

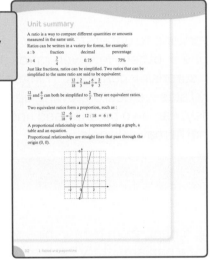

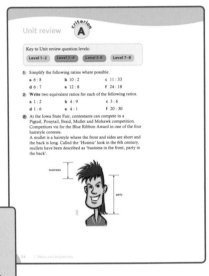

Unit review allows you to practice the skills and concepts in the unit, organized by level (criterion A) as you might find on a test or exam. You can get an idea of what achievement level you are working at based on how well you are able to answer the questions at each level. *Answers for these questions are included at the back of the book.*

The **summative assessment** task applies the mathematics learned in the unit to further explore the global context. This task is often assessed with criteria C and D.

1 Ratios and proportions

In this unit, you will use ratios and proportions to understand how humans compete and cooperate with one another. However, ratios and proportions can also be useful in other, surprising contexts. Things that we often take for granted, like the fairy tales we heard growing up or the politics that we hear about every day, can also be analysed and understood through these important mathematical tools.

Fairness and development

Politics and government

There are many different forms of government, from monarchies to democracies to republics. Democratic countries may use a parliamentary system of government or a presidential one. What do these terms mean?

Is one system more representative of the citizens than another? Do the different systems represent a country's people equally, despite their differences? What is the ratio of representatives to citizens? Is this ratio the same in urban areas and rural areas? Is it possible that some people have no representation at all?

Personal and cultural expression

Does magic know math?

Traditional tales are a part of growing up all over the world. What if those stories were actually true? Is that possible? A look at popular fables through the lens of ratios and proportions might reveal some flaws in their logic. Would the morals of these stories be more believable if the stories were as well?

Hansel and Gretel is a story written by the Brothers Grimm about a young brother and sister. Alone in the woods, they encounter a witch who entices them to come with her, all in the hope of eating them!

The witch's cottage is made of gingerbread and candy, which tempts the children in. How would a typical gingerbread recipe have to be adapted to create a cottage large enough to live in? How much would those ingredients cost?

Cinderella tells the story of a girl who meets a prince at a ball. She has to leave before her carriage turns back into a pumpkin and her horses into mice and, in the rush, she leaves behind a glass slipper. The prince visits every girl in the kingdom to find the owner of the slipper. Could he have used proportions to narrow down his search?

When the fairy godmother transforms a pumpkin into a carriage and mice into horses, are they all enlarged using the same ratio? Does magic know math or is it really just luck?

1 Ratios and proportions
Competition and cooperation

Related concepts: Equivalence, Quantity and Simplification

Global context:

Part of studying the global context **identities and relationships** involves looking at how humans cooperate and compete with one another. Ratios and proportions allow you to analyze these human interactions and even make decisions about what may be fair or desirable. There are also times when competitors are forced to cooperate with one another, blurring the lines between who may be a friend and who may be a foe.

Statement of Inquiry:

Using a logical process to simplify quantities and establish equivalence can help analyse competition and cooperation.

Objectives

- Defining and simplifying ratios
- Dividing a quantity in a given ratio
- Defining proportion and demonstrating proportional relationships
- Representing proportional relationships using tables, equations and graphs
- Finding the constant of proportionality for a proportional relationship
- Applying mathematical strategies to solve problems using proportional reasoning

Inquiry questions

F
What does it mean to simplify?
What does it mean to be equivalent?

C
How can you establish equivalence?
How are simplification and equivalence related?

D
What makes for healthy and fair competition?
Which is more about being equal, competition or cooperation? Explain.

ATL1 Organization skills

Create plans to prepare for summative assessments (examinations and performances)

ATL2 Affective skills

Practise positive thinking

You should already know how to:

- round numbers correctly
- solve problems involving percentages
- convert between different units
- solve simple equations

Introducing ratio and proportion

Competition among humans is a very natural part of who we are. There are professional sports teams who compete with each other, as well as the adolescents who are hoping to be the next great athlete. The Olympics bring together athletes from around the world for two weeks of competition. There is also the day-to-day competition among people seeking that exciting new job.

Despite this constant competition, humans must also cooperate with one another. Those same athletes and those same employees must also find a way to work *with* others to accomplish group goals.

Where is the line between competition and cooperation? Can it be analysed? In this unit, you will learn to describe competition and cooperation among humans and, hopefully, understand where that line is.

Reflect and discuss 1

In a small group, answer the following:

- Give two examples of times when you were competitive.
- Give two examples of times when you were cooperative.
- Which do you experience more, competition or cooperation? Explain.
- Has there ever been a time when you competed against someone with whom you later had to cooperate? If so, describe how that went.

Ratios

In mathematics, a *ratio* is a way to compare different quantities that are measured using the same units. Just like fractions, ratios can be simplified.

For example, if there are 12 girls and 8 boys on your team, the ratio of girls to boys would be written as 12 to 8 or 12 : 8 (also read as '12 to 8').

In simplified form, this would be written as 3 to 2 or 3 : 2. Because the units are the same for both quantities (people), we do not need to write them.

The ratio of girls to the total number of students would be written as 12 to 20, or 12 : 20. In simplified form, this would be written as 3 : 5.

The ratio of boys to girls to the total number of students would be written as 8 to 12 to 20, or 8 : 12 : 20. In simplified form, this would be written as 2 : 3 : 5.

 ## Investigation 1 – Simplifying ratios

a Copy the table below.

Unsimplifed ratio	Simplified ratio
8 : 12	
21 : 35	
18 : 9	
4 : 16	
15 : 35	
45 : 10	
36 : 42	

1 : 4   9 : 2   2 : 3   7 : 3

3 : 7   6 : 7   3 : 5

5 : 6   2 : 1   1 : 9

8 : 3   6 : 5   5 : 4

b Find the simplified form of each ratio in the table from among the options on the right. Write them in the table next to the unsimplified ratio.

c Explain how you figured out which ratios belong together.

d Generalize the procedure to simplify a ratio.

e Verify your procedure with two examples different than the ones above.

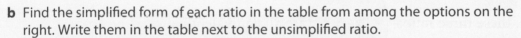

Reflect and discuss 2

- Can all ratios be simplified? Explain.
- How is simplifying ratios similar to simplifying fractions? Explain.

Activity 1 – Ratios around the room

Pairs

1 With your partner, identify some ratios that you see around your classroom. Remember that a ratio is a comparison of two quantities with the same units.

2 With your partner, play **I Spy**. The first player looks around the room and finds two quantities that can be expressed as a ratio. The player says, "I spy with my little eye a ratio of _____ ", filling in the blank with a ratio. The second player tries to guess what the two quantities represent. When the second player is successful, the roles are switched.

3 Play **I Spy** four times each, with the last round using ratios that have to be simplified.

Reflect and discuss 3

In pairs, answer the following questions:

- Can you make a ratio out of any two quantities? Explain.

- Is a ratio of 1 : 2 the same as a ratio of 2 : 1? Explain.

Two ratios that can be simplified to the same ratio are said to be *equivalent*.

For example, these three ratios: (12 : 16   3 : 4   21 : 28) are all equivalent

because they all simplify to 3 : 4.

Investigation 2 – Equivalent ratios

criterion B

1 Copy the following table.

Ratio	Equivalent ratios				Simplified ratio
4 : 6	40 : 60	12 : 18	24 : 36	36 : 54	
30 : 12	60 : 24	15 : 6	300 : 120	120 : 48	
10 : 40	5 : 20	30 : 120	80 : 320	2 : 8	
8 : 28	16 : 56	80 : 280	32 : 112	40 : 140	
16 : 8	8 : 4	48 : 24	4 : 2	320 : 160	
14 : 20					
36 : 6					

▶ Continued on next page

2 Find the simplified form of each ratio and write it in the right-most column.

3 Write down how each equivalent ratio was found, given the ratio in the left-most column.

4 In the middle column, write down another ratio equivalent to those given.

5 Complete the last two rows of the table by writing in four equivalent ratios each.

6 Write down how to find a ratio that is equivalent to a given ratio.

7 Verify your procedure with two examples that are different than the ones in the table.

Reflect and discuss 4

- How are equivalent ratios like equivalent fractions? Explain.

- Knowing that ratios generally contain whole numbers, how would you simplify the ratio 0.5 : 3?

Example 1

 One of the most famous competitions in the 1970s was the 'Battle of the Sexes' tennis match between Billie Jean King and former tennis champion Bobby Riggs.
Riggs challenged all female tennis players to defeat him, which King did in 1973. The ratio of games won was 18 : 10 in favor of King. Give three ratios equivalent to 18 : 10.

A 18 : 10 = 9 : 5

> Divide each quantity by 2.
> Note that this is now a simplified ratio since neither quantity can be divided further.

18 : 10 = 36 : 20

> Multiply each quantity in the original ratio by any whole number, in this case 2....

18 : 10 = 90 : 50

> ...and in this case, 5.

Three ratios equivalent to 18 : 10 are 9 : 5, 36 : 20 and 90 : 50.

Equivalent ratios can be very helpful when trying to solve a wide range of problems involved in both cooperation and competition.

Example 2

Q A marathon is a race of approximately 42 km. In many marathon competitions, there is a relay category in which a team of runners works together to complete the race, with each competitor running some portion of the entire distance.

a Suppose three friends create a relay team and, because of their differing abilities, they decide to run distances in a ratio of $1 : 2 : 3$. How far does each person run?

b After the first 2 km, the first runner injures himself. The remaining two runners decide to run the rest of the race in a ratio of $3 : 5$. How far did the remaining two competitors run?

A **a** $1 + 2 + 3 = 6$

> Find the total of the number of terms in the ratio.

Runner 1's portion $= \dfrac{1}{6}$

> Express each runner's contribution as a fraction of the total.

Runner 2's portion $= \dfrac{2}{6}$ or $\dfrac{1}{3}$

Runner 3's portion $= \dfrac{3}{6}$ or $\dfrac{1}{2}$

Runner 1's portion $= \dfrac{1}{6}$ of 42, or $\dfrac{1}{6} \times 42 = 7$

> Calculate the fraction of the race run by each runner.

Runner 2's portion $= \dfrac{1}{3}$ of 42, or $\dfrac{1}{3} \times 42 = 14$

Runner 3's portion $= \dfrac{1}{2}$ of 42, or $\dfrac{1}{2} \times 42 = 21$

$7 \text{ km} + 14 \text{ km} + 21 \text{ km} = 42 \text{ km} \checkmark$

> Check your solution to make sure it works.

One runner will run 7 km, another 14 km and the third one 21 km.

b Find a ratio equivalent to $3 : 5$

> At the 2 km mark, there are 40 km remaining in the race. The goal is to find a ratio equivalent to $3 : 5$ where the values sum to 40.

$3r : 5r$
$3r + 5r = 40$

> Thus, if you multiply each of the values by the same constant, then that should sum to 40.

$8r = 40$
$r = 5$

> Simplify the left-hand side and solve.

▶ Continued on next page

$3r = 15$ km
$5r = 25$ km

15 km $+ 25$ km $= 40$ km ✓

One runner will run 15 km and the other will run 25 km.

Substitute the value for r back into the ratio $3r : 5r$ to find the distance each person ran.

Check your solution to make sure it works.

Did you know?

The legend goes that Pheidippides, a Greek messenger, ran 25 miles from Marathon to Athens, to deliver the news of the Athenians' victory over the Persians in 490 B.C. The modern marathon distance became 26.2 miles at the 1908 Summer Olympics in London, where the marathon course was designed so that it could start at Windsor Castle and finish at the Olympic stadium.

The oldest marathon in the United States is the Boston Marathon, which has been run continually since 1897.

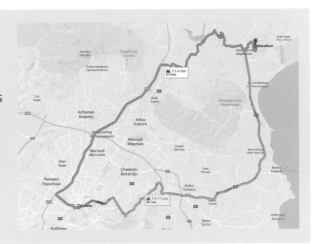

Practice 1

1 Simplify the following ratios. If a ratio is already in its simplest form, write the word 'simplified'.

a $3 : 5$	**b** $4 : 6$	**c** $12 : 16$
d $100 : 200$	**e** $700 : 35$	**f** $0.4 : 6$
g $9 : 10$	**h** $1 : 6$	**i** $950 : 50$
j $13 : 200$	**k** $1006 : 988$	**l** $64 : 16$

2 Write three equivalent ratios for each of these ratios:

a $2 : 5$	**b** $2 : 18$	**c** $9 : 27$	**d** $10 : 25$
e $6 : 7$	**f** $15 : 45$	**g** $8 : 1$	**h** $12 : 3$

3 DC and Marvel created comics with teams of superheroes in order to compete for readers. DC had 'The Justice League' while Marvel had 'The Avengers'. The ratio of the number of original members for the respective superhero teams was (DC to Marvel) 7 : 5.

▶ Continued on next page

a Over a period of several years, each company added new superheroes to their team. If the ratio of the number of superheroes was kept the same over that period, give two examples for the number of superheroes that could have been in each team.

b Some time later, the ratio of superheroes (DC to Marvel) was 21 : 10. Is this ratio equivalent to the ratio in part **a**? Show your working.

c At one point, The Justice League had 91 superhero members. How many superheroes would you expect the Avengers to have if the original ratio was maintained? Show your working.

4 Draw a picture that includes all of the following elements:

- a 2 : 3 : 5 ratio of green flowers to blue flowers to red flowers
- boys and girls in a ratio of 1 to 2
- trees and bushes in a ratio of 1 : 1.

You may add whatever else you like to your picture to make it unique.

5 From the following list, write down all of the ratios that are equivalent to each other.

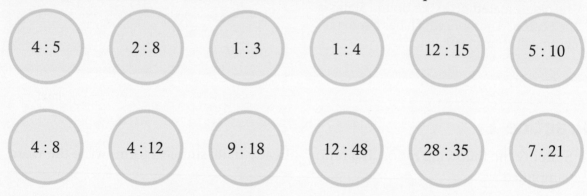

6 Use the diagram here to justify why the ratio 3 : 4 is equivalent to the ratio 15 : 20.

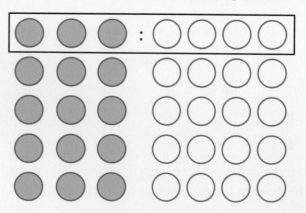

▶ Continued on next page

7 In the Pokémon card game, two players enter into friendly competition to 'battle' each other in a field and knock out each other's Pokémon. A starting deck of cards could have a ratio of Pokémon cards to trainer cards to energy cards of 5 : 2 : 3. If the deck contains 60 cards, how many of each card type will there be?

8 Men and women often do the same job, though they do not always get the same pay. When men and women get different pay for the same work, there is a *gender pay gap*. The ratio of women's wages to men's wages for four countries is given in this table.

Country	Women's wages : Men's wages for same work (currency)	Simplified ratio
Bulgaria	1.5 : 2 (lev)	
Australia	0.83 : 1 (dollar)	
Russia	2.40 : 3 (ruble)	
United Arab Emirates	0.99 : 1 (dirham)	

a Complete the table by finding a simplified ratio for each country.

b Which country has the largest gender pay gap? Explain.

c In which country does the gender pay gap disadvantage women the most? In which country does the gender pay gap disadvantage women the least? Explain.

9 Given the following ratios and totals, find the portion represented by each part of the ratio.

a ratio = 3 : 7 , total = 90   **b** ratio = 1 : 2 : 5 , total = 88

c ratio = 9 : 3 : 4 , total = 32   **d** ratio = 2 : 4 : 5 : 9 , total = 120

10 In a regional mathematics competition, the ratio of Swedish students to Danish students to Finnish students is 14 : 22 : 13. Altogether, there are 1029 students. Find how many students come from each of the three countries. Show your working.

11 Given that the ratios in each part below are equivalent, find what quantity each variable represents.

a $1 : 4 = 2 : k$ **b** $2 : 5 : 8 = 16 : p : 64$ **c** $8 : 7 = 24 : x$

d $2 : 3 : 7 = 14 : r : s$ **e** $2 : 7 : 10 = 1 : q : 5$ **f** $4 : 1 : 12 = 6 : z : 18$

Reflect and discuss 5

- In terms of your learning, write down what is going well so far in this unit.

- Describe a time when thinking positive thoughts made a difference.

Other ways to represent ratios

Ratios can be represented in a variety of forms and you should feel comfortable moving between these different representations. How to do that is the focus of the next activity.

Activity 2 – Competition in government

Becoming an elected official usually involves competing against other men and women for the same position. Once elected, these officials must then cooperate with people from different political parties and viewpoints in order to effect change. In this activity, you will explore the ratio of women to men that hold elected positions in the governments of various countries.

Below are data showing the number of elected seats held by men and women around the world.

Country	Women	Men	Total elected officials	Ratio of women to men
Argentina	130	199	329	
Australia	73	153	226	
Canada	88	250	338	
France	223	354	577	
Sweden	152	197	349	
United Kingdom	191	459	650	
United States	104	431	535	

1 Make a copy of the table and write the ratio of women in government to men in government for the seven countries.

2 Which country has the highest ratio of women to men? Which has the lowest? How can you tell?

3 Now, you are going to use a decimal and a percentage to help compare the quantity of women in government positions around the world. Fill in a table like the following to help organize your information.

▶ Continued on next page

Country	Ratio of women out of **total members** of government expressed as a fraction	Ratio of women out of total members of government expressed as a decimal	Ratio expressed as a percentage (of women out of total members of government)
Argentina			
Australia			
Canada			
France			
Sweden			
United Kingdom			
United States			

4 What does a percentage mean as a ratio? Explain.

5 What are the advantages of using percentages to represent ratios?

6 In 1991, Argentina adopted the world's first gender quota law. The law mandates that political parties nominate women for at least 30% of the electable positions on their candidate list. Which of the countries above have at least 30% of women in government?

Reflect and discuss 6

- Explain a general rule for converting a ratio to a percentage.
- Would it be possible to represent the ratio 3 : 5 : 11 as a fraction or a percentage? Explain.
- Describe the advantages and disadvantages of representing a ratio as a fraction and as a percentage.

Practice 2

1 Match each ratio card with a fraction card **and** a percentage card.

Ratio cards

112 : 564	42 : 70
24 : 30	23 : 46
2.5 : 6	17.5 : 21
8 : 72	18 : 54

Fraction cards

$\frac{5}{12}$	$\frac{4}{5}$
$\frac{3}{5}$	$\frac{1}{2}$
$\frac{1}{3}$	$\frac{28}{141}$
$\frac{5}{6}$	$\frac{1}{9}$

Percentage cards

80%	19.9%
50%	41.7%
83.3%	11.1%
33.3%	60%

2 Universities and colleges around the world use a selective admissions process to build a diverse population of students. Students of all backgrounds compete with one another for the limited number of places in these institutions of higher education.

For a top university in North America, these are approximate ratios for the number of students admitted from a variety of groups compared with the whole school population.

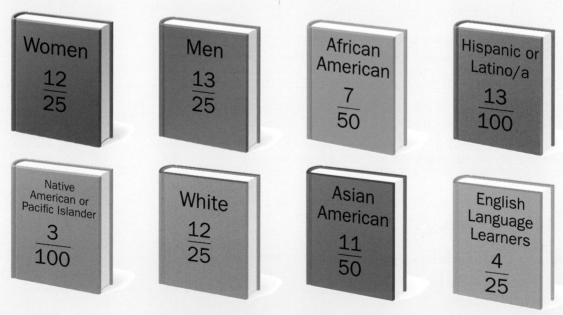

Women $\frac{12}{25}$ Men $\frac{13}{25}$ African American $\frac{7}{50}$ Hispanic or Latino/a $\frac{13}{100}$

Native American or Pacific Islander $\frac{3}{100}$ White $\frac{12}{25}$ Asian American $\frac{11}{50}$ English Language Learners $\frac{4}{25}$

a What do these fractions mean? Explain using an example.

b Represent these quantities as a ratio, decimal and percentage. Which form is most effective to represent the diversity of the school? Explain.

▶ Continued on next page

c Show that the number of women and men admitted to the university is approximately equal. Explain which representation allows you to see this most easily.

3 The Summit Series was a famous hockey competition between Canada and the USSR during the Cold War. In 1972 and again in 1974, the Soviet team played against Canadian all-stars from the National Hockey League in an 8-game series. Half of the games were played in Canada and half in Moscow. Players on Team Canada, who would normally play against one another, now joined forces to represent their country.

The results of the 1972 series are given below.

Played in Canada

Game	Score
1	USSR 7 – Canada 3
2	Canada 4 – USSR 1
3	Canada 4 – USSR 4
4	USSR 5 – Canada 3

Played in the Soviet Union

Game	Score
1	USSR 5 – Canada 4
2	Canada 3 – USSR 2
3	Canada 4 – USSR 3
4	Canada 6 – USSR 5

a Write the ratio of games won to total number of games played for each team. Represent the quantities in four different ways (ratio, fraction, decimal, percentage).

b Which representation is the most effective when trying to describe who won the series? Explain.

c Write the ratio of goals scored by the Soviet Union to the number of goals scored by Team Canada. Express this ratio in four different ways.

d Which representation in part c is most effective? Explain.

▶ Continued on next page

4 The International Space Station (ISS) is a habitable station orbiting Earth several times each day. It is a joint project of five space agencies from Europe, Canada, Japan, Russia and the United States and it has been continuously inhabited since the year 2000. Its goal is to help conduct scientific research and possibly be a stopping point for future trips to the Moon or Mars.

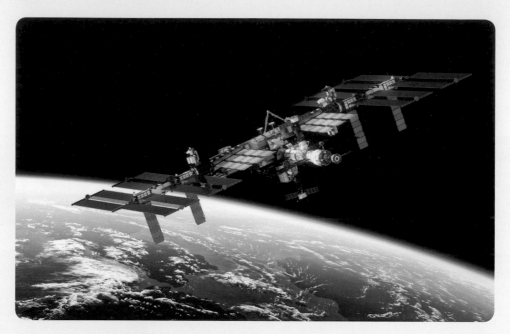

 a Go online to find the current astronauts aboard the ISS. Represent the ratio of the number of astronauts from each country in four different ways. Which representation is the most effective? Explain.

b Expedition 42 had a ratio of women to men of 2 : 4. Represent the ratio of women to the total number of astronauts as a percentage.

c It is very expensive to bring water to the station, so much of the water consumed by astronauts is recycled from a variety of sources. The ratio of the amount of water used for food (for one astronaut) to the amount of water one astronaut drinks is 21 : 43. Find the percentage of total water that is used for drinking (assuming that water is not used for any other purposes).

d Because the ISS orbits the Earth several times each day, astronauts see multiple sunrises and sunsets. An "ISS day" is 6.25% of an Earth day. Represent the ratio of an ISS day to an Earth day as a fraction in its simplest form. Hence find the number of times the ISS orbits Earth in one day.

Formative assessment

Humans have created competitions in a wide range of activities. Some involve time, some involve measurements and others are judged. Every summer since 1972, participants from around the world have competed in a hotdog-eating contest on Coney Island, New York. In this event, competitors have to eat as many hotdogs as they can in a given time.

In 2006, men and women competed against each other. The competition was 12 minutes long and produced the results in Table 1.

Cash prizes were awarded beginning in 2007. The male and female winners each receive US$10 000.

	Position	Competitor	Hotdogs eaten	Competitor's weight (kg)
Table 1	1	Takeru Kobayashi	53.75	58
	2	Joey Chestnut	52	104
	3	Sonya Thomas	37	44

In 2017, competition was based on gender and lasted only 10 minutes. The men's results are in Table 2.

	Position	Competitor	Hotdogs eaten	Prize money (US$)
Table 2	1	Joey Chestnut	72	10 000
	2	Carmen Cincotti	60	5 000
	3	Matt Stonie	48	2 500

▶ Continued on next page

The 2017 women's results are shown in Table 3.

	Position	Competitor	Hotdogs eaten	Prize money (US$)
Table 3	1	Miki Sudo	41	10 000
	2	Michelle Lesco	32.5	5 000
	3	Sonya Thomas	30	2 500

a Write down the ratio of prize money to the number of hotdogs eaten for each of the top three male and female competitors in 2017.

b Simplify this ratio and then express it in four different ways (a : b, fraction, decimal, percentage).

c Which representation is most useful? Explain.

d Represent the ratio of the number of hot dogs eaten by each of the winners (from all of the competitions) to the duration of the competition in four ways.

e Which representation in part **d** makes Joey Chestnut's world record in 2017 most visible? Explain.

f Use a ratio to show that the results of the 2006 contest could have been different. Make sure that your ratio **clearly** demonstrates why the results could have been different. Would this be a fair way to determine the winner that could include both genders? Justify your answer.

g Is this competition more fair now that the contest is divided by gender? Explain.

ATL2 Reflect and discuss 7

- Write down what is going well so far in this unit in terms of your learning.

- Describe something positive about yourself that you know you can always rely on in class.

Proportions

Two equivalent ratios form a *proportion*, such as the following:

$$\frac{2}{15} = \frac{6}{30} \quad \text{or} \quad 2:15 = 6:30$$

Proportions and proportional reasoning can be extremely helpful in solving problems.

Solving proportions

You have previously worked with equivalent fractions. An equation showing that two fractions or ratios are equivalent is called a proportion. These proportions contain patterns that you will discover in the next investigation.

Investigation 3 – Solving proportions

Look at the following equivalent fractions/proportions.

$$\frac{2}{3} = \frac{4}{6} \qquad \frac{10}{20} = \frac{4}{8} \qquad \frac{1}{6} = \frac{4}{24} \qquad \frac{1}{4} = \frac{10}{40}$$

$$\frac{10}{15} = \frac{2}{3} \qquad \frac{5}{6} = \frac{10}{12} \qquad \frac{3}{4} = \frac{6}{8} \qquad \frac{2}{5} = \frac{6}{15}$$

$$\frac{3}{8} = \frac{12}{32} \qquad \frac{2}{7} = \frac{6}{21} \qquad \frac{4}{9} = \frac{20}{45} \qquad \frac{100}{250} = \frac{4}{10}$$

criterion **B**

a Write down patterns you see that are the same for **all** pairs.

b Share the patterns you found with a peer. Which one(s) can be written as an equation?

c Generalize your patterns by writing one or more equations based on the proportion $\frac{a}{b} = \frac{c}{d}$.

d Add four more equivalent fractions/proportions to the ones here and verify your equations for these new proportions.

Reflect and discuss 8

- How could you use your equation(s) to find the missing value in each of the following?

  $$\frac{3}{8} = \frac{x}{96} \qquad \frac{m}{11} = \frac{75}{165} \qquad \frac{2.4}{x} = \frac{8.4}{20.5}$$

- Explain how converting a ratio represented as a fraction to a percentage involves a proportion.

Example 3

Q Find the missing value in each of these proportions. Round your answers to the nearest hundredth where necessary.

a $\dfrac{8}{22} = \dfrac{12}{x}$

b $\dfrac{5}{9} = \dfrac{x}{17}$

 A **a** If $\dfrac{8}{22} = \dfrac{12}{x}$,

then $8x = 12 \times 22$

In a proportion, the products of the quantities on each diagonal are equivalent.

$8x = 264$

Simplify the right-hand side

$\dfrac{8x}{8} = \dfrac{264}{8}$

Solve the equation by dividing each side by 8.

$x = 33$

The missing value is 33.

b If $\dfrac{5}{9} = \dfrac{x}{17}$,

then

$9x = 5 \times 17$

In a proportion, the products of the quantities on each diagonal are equivalent.

$9x = 85$

Simplify the right-hand side

$\dfrac{9x}{9} = \dfrac{85}{9}$

Solve the equation by dividing each side by 9.

$x = 9.\dot{4}$

The missing value is approximately 9.44.

Activity 3 – Solving proportions: a different approach

Another way to solve proportions uses the techniques you learned to solve equations.

1 Solve the following equations.

$$\frac{w}{4} = 5 \qquad 10 = \frac{y}{3} \qquad \frac{x}{3} = \frac{21}{9} \qquad \frac{5}{8} = \frac{h}{16}$$

2 Explain how you solved them. Justify your process.

3 Given the equivalent fractions below, find the reciprocal of each.

$$\frac{5}{10} = \frac{1}{2} \qquad \frac{3}{4} = \frac{12}{16} \qquad \frac{24}{18} = \frac{8}{6} \qquad \frac{10}{35} = \frac{2}{7}$$

4 What do you notice about your answers? Are the new fractions still equivalent to each other? Explain.

5 Use your result from part **4** in order to solve the following equation.

$$\frac{10}{7} = \frac{21}{x}$$

6 Describe how to solve proportions using this method.

7 Describe how this method relates to the method learned in Investigation 3.

There are, in fact, many ways of solving proportions. In future classes, you may encounter other methods. It is important to understand how all of these methods relate to the fundamental mathematical principles that you already know.

WEB LINK

Search for "Exploring rate, ratio and proportion" on the LearnAlberta.Ca website. In this interactive activity, you can practice equivalent 3 term ratios and watch a video and answer questions on ratios in photography.

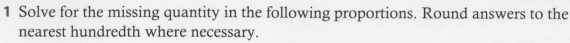

Practice 3

1 Solve for the missing quantity in the following proportions. Round answers to the nearest hundredth where necessary.

a $1 : 4 = 2 : w$

b $8 : 7 = 24 : x$

c $p : 12 = 5 : 4$

d $\dfrac{3}{8} = \dfrac{m}{24}$

e $\dfrac{h}{57} = \dfrac{11}{19}$

f $\dfrac{8}{15} = \dfrac{19}{x}$

g $\dfrac{26}{81} = \dfrac{a}{100}$

h $\dfrac{31}{b} = \dfrac{15}{27}$

i $\dfrac{3}{7} = \dfrac{10}{y}$

j $\dfrac{12}{w} = \dfrac{5}{8}$

k $\dfrac{v}{21} = \dfrac{7}{10}$

l $\dfrac{5}{z} = \dfrac{45}{7}$

m $\dfrac{1.2}{6.8} = \dfrac{3.4}{c}$

n $\dfrac{4}{7.3} = \dfrac{f}{2.1}$

o $\dfrac{k}{43.5} = \dfrac{11.2}{87}$

p $\dfrac{5.5}{8} = \dfrac{18}{y}$

2 Determine whether each of the following pairs of ratios are equivalent. Justify your answer using the rule you discovered.

a $\dfrac{3}{4}$ and $\dfrac{42}{56}$

b $\dfrac{8}{11}$ and $\dfrac{21}{29}$

c $4 : 15$ and $\dfrac{21}{32}$

d $13 : 5$ and $32.5 : 13.5$

3 Orienteering competitions combine navigation with speed, where participants use a map to help navigate their route. Maps use a scale to represent distances in real-life, such as the one below.

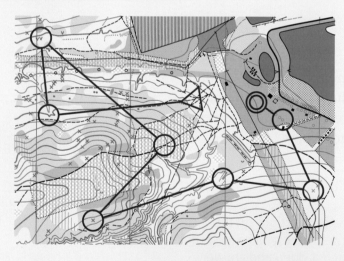

The scale of this map is 1 : 15 000, meaning that 1 unit measured on the map is equivalent to 15 000 of the same unit measured in real life. In this example, 1 cm on the map is equivalent to 15 000 cm (or 150 m) in real life.

a If you measured 4.7 cm on the map, how far is that in real life (in meters)? Show your work using a proportion.

b How far would a distance of 350 m in real life measure on the map? Show your work using a proportion.

c On the map, the distance between two control points (circles) is 79 mm. Find the distance in real life (in meters) using a proportion.

▶ Continued on next page

4 The Great Trail is the longest continuous recreational trail system in the world. Spanning the ten provinces and three territories of Canada, the 24 000 km trail was created through the cooperation of 477 organizations and thousands of volunteers.

a The trail took 25 years to complete. Assuming a proportional relationship, find how much of the trail would have been finished after the first 10 years. Round your answer to the nearest kilometer.

b After how many years would 80% of the trail have been finished? Use a proportion to find your answer and round it to the nearest year.

c Assuming each organization was responsible for an equivalent amount of the trail, how many organizations would have been necessary to create the first 15 000 km of trail? Round your answer to the nearest whole number.

d Canada has a population of approximately 37 million people. How many (equally sized) people would occupy one kilometer if they were to all take a position along the Great Trail? Show your working using a proportion.

5 Climbing a mountain like Mt Everest requires incredible teamwork. Not only are the conditions difficult, but the equipment necessary to safely scale the mountain can be very heavy. With a trek of about two weeks from Kathmandu to Base Camp, it is imperative that humans cooperate if everyone is to make it back safely. Groups generally hire Sherpa porters to help them with carrying the heavy equipment. If a group with 540 kg of equipment requires 12 Sherpas, how many Sherpas should be required to carry 945 kg? Show your working.

For more practice with ratios and solving proportions, search for "Dunk Tank! Ratio & Proportion" on the PBS learning media website. In this interactive game, you will use ratios and proportions to solve problems in a variety of activities.

Recognizing and using proportional reasoning

As you have seen, proportions are powerful tools for solving problems. However, how do you know when a relationship is proportional? Can you solve every problem with proportions, or do only certain kinds of questions involve *proportional reasoning*?

Investigation 4 – Recognizing proportional relationships

Many countries provide incentives for their athletes at the Olympics. While British athletes, for example, receive no monetary award for winning medals, some countries, such as India, pay their athletes for outstanding performances in order to increase their competitive spirit.

In India, the amount of money earned per gold medal is given below. This is a *proportional relationship*.

Number of gold medals	Payment (euros)
2	200 000
3	300 000
4	400 000
5	500 000

a Plot the number of gold medals on the *x*-axis and the payment on the *y*-axis. Describe the type of graph obtained.

b How much would a person be paid for winning *no* gold medals? How much would a person be paid for winning 10 gold medals? Explain how you found your answers.

In Thailand, gold medallists earn US$314 000 to be paid in equal amounts over 20 years.

c How much would an athlete be paid each year? Show your working.

d Fill in the table on the right and graph your results.

e Is this a proportional relationship? Explain.

f What do you multiply by to find the total money paid in any year? This is called the *constant of proportionality*.

g Find an equation for your graph. Explain how the constant of proportionality is represented in the equation.

Year	Total money paid (US$)
0	
1	
2	
3	
4	
5	
10	
20	

▶ Continued on next page

h Explain why each example below does not represent a proportional relationship.

i
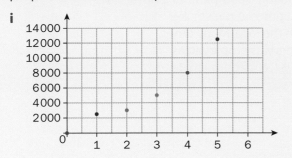

ii
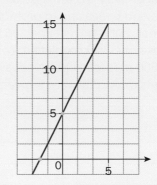

iii

X	Y
2	10
5	25
8	30
10	45
12	60

iv

X	Y
0	10
2	20
4	40
6	60
8	80

v

X	Y
1	10
2	20
3	30
4	20
5	10

i Write down a rule for determining whether a relationship is proportional, given data in a table.

j Write down a rule for determining whether a relationship is proportional, given its graph.

k Write down a rule for determining whether a relationship is proportional, given its equation.

Reflect and discuss 9

- What makes a relationship a *proportional relationship*? Explain.
- Does a proportional relationship have to include the point (0, 0)? Explain.
- Explain how the graph, equation and table all represent the same characteristics of a proportional relationship.

A *proportional relationship* can be represented using a graph, a table and an equation. Proportional relationships exist when one variable can be obtained by multiplying the other by a constant. This multiplicative factor is called the *constant of proportionality*. Once you have recognized a relationship as being a proportion, then you can use proportional reasoning to solve problems involving that relationship.

Activity 4 – Keeping the pace

A runner called a pacemaker is often used in races to force competitors to try their hardest the entire race and to promote fast times. A pacemaker is often a good runner, but is never expected to win the race. Many times, the pacemaker does not even complete it. Tom Byers was employed as a pacemaker in a 1500 m race in Oslo, Norway in 1981. His approximate time at various points during the race is given below.

Distance (m)	Elapsed time (s)
100	14.5
200	29
400	58
800	116

1 Does this seem like a proportional relationship? Explain.

2 Draw the graph of Byers' time versus the distance he covered.

3 Find an equation for your graph. Verify that it works by substituting in at least one of the points.

4 Is this a proportional relationship? Justify your answer.

5 Find Byers' time for the 1500 m race based on your findings. Show how the same result is achieved using the equation, the graph and a proportion.

6 Byers' final time was actually 219.01 seconds. Explain why this is different than the value you found.

7 Does this mean this is not a proportional relationship? Explain.

Did you know?

Steve Ovett and Steve Cram were the favorites to win the race, but they chose not to keep up with Byers' fast pace. At one point, Byers had a lead of over 70 m. Ovett, who held the world record at the time, finished the race in second place, 0.5 seconds behind Byers!

STEVE OVETT: OLYMPIAN

Practice 4

1 Determine whether or not each of the following is a proportional relationship. If so, find its equation.

a

X	Y
0	20
1	30
2	40
3	50

b

X	Y
3	12
5	20
7	28
15	60

c

X	Y
0	0
1	2
2	5
3	10

d

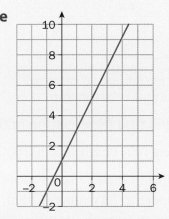

e

f

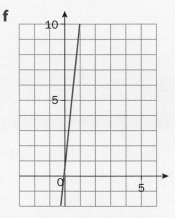

2 The European Economic and Social Committee (EESC) has a video and singing competition where groups of singers create a video of themselves singing a song chosen by the EESC. A group of six singers sang the required song in 4 minutes. Another group of 12 vocalists will sing the same song. How long will it take this group to sing it?

3 Rideshare companies are becoming a common mode of transportation. In order to compete with current pricing strategies, a new company called *Wheelz* is going to charge customers $2 for every 3 minutes of driving time. How long will a 7-minute ride cost?

4 This graph represents the money earned ($US) per gold medal at the Olympics, as promised by the United States Olympic Committee (USOC).

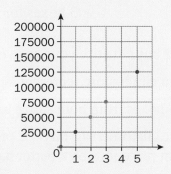

▶ Continued on next page

a Is this relationship proportional? Explain.

b What is the constant of proportionality?

c How much money would someone who won eight gold medals receive? Show your working.

5 In 1989, two million people held hands along the side of the road through Estonia, Latvia and Lithuania. Named 'the Baltic Way', the people cooperated to put on a peaceful protest in favor of independence that achieved positive results.

Number of people	Length of chain of hands (m)
20	6.4
100	32
500	160
1200	384

a Show that this is a proportional relationship.

b Draw a graph of the data and find the equation of the line.

c Find the length in kilometers of a chain of 2 000 000 people.

6 Grocery stores often price items in bulk to seem more competitive with other stores, but they also encourage shoppers to buy more. Often, you can get the same price per item and buy fewer of them, even though the advertising might lead you to believe otherwise. The following are prices advertised by two different stores, though you are allowed to buy more or less than the amount indicated.

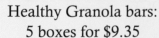

Healthy Granola bars: Healthy Granola bars:
5 boxes for $9.35   3 boxes for $5.58

a Which store has the better buy? Justify your answer.

b Does buying this item from either store represent a proportional relationship? Explain.

c Find the cost of buying 15 boxes from both stores using an equation and a proportion.

d If you were to try to entice customers to buy the product from your store, give an example of how you might price it to be a better bargain, but also entice shoppers to buy more than one.

Reflect and discuss 10

ATL1

- Write down what went well in this unit in terms of your learning.
- Was there anything in this unit that you didn't understand at first, but now you do? How did you overcome that challenge?

ATL2

- You are now capable of doing your summative task. Write down a plan to complete each section before the due date.
- Create a mind map to summarize what you have learned in this unit. Compare with a peer to make sure you each have all of the concepts.
- Write down an example of how to do each kind of problem that covers a particular concept that you have learned in this unit.
- Create a daily plan to study for your unit test.

Unit summary

A ratio is a way to compare different quantities or amounts measured in the same unit.

Ratios can be written in a variety for forms, for example:

a : b	fraction	decimal	percentage
3 : 4	$\dfrac{3}{4}$	0.75	75%

Just like fractions, ratios can be simplified. Two ratios that can be simplified to the same ratio are said to be equivalent:

$$\frac{12}{18} = \frac{2}{3} \text{ and } \frac{6}{9} = \frac{2}{3}$$

$\dfrac{12}{18}$ and $\dfrac{6}{9}$ can both be simplified to $\dfrac{2}{3}$. They are equivalent ratios.

Two equivalent ratios form a proportion, such as :

$$\frac{12}{18} = \frac{6}{9} \quad \text{or} \quad 12 : 18 = 6 : 9$$

A proportional relationship can be represented using a graph, a table and an equation.

Proportional relationships are straight lines that pass through the origin (0, 0).

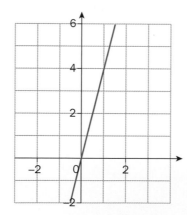

The equation of a proportional relationship always has the form $y = kx$, where k is called the constant of proportionality.

x	y
2	6
3	9
5	15
12	36

The table of a proportional relationship can be recognized by the fact that the y-coordinate (dependent variable) is always the same multiple of the x-coordinate (independent variable). This multiplicative factor is the constant of proportionality.

Solving a proportion can be done in a variety of ways.

The products on the diagonals are equal, which produces an equation that can be solved.

$$\frac{12}{14} = \frac{18}{x} \qquad 12x = 14 \times 18$$

You can also solve it like an equation. If the variable is in the denominator, simply take the reciprocal of each ratio and then solve the equation.

$$\frac{d}{5} = \frac{6}{15} \qquad\qquad \frac{5}{d} = \frac{15}{6}$$

$$5 \times \frac{d}{5} = \frac{6}{15} \times 5 \qquad \frac{d}{5} = \frac{6}{15}$$

$$d = 2$$

Unit review

criterion **A**

📖 **Launch additional digital resources for this chapter**

Key to Unit review question levels:

Level 1–2   Level 3–4   Level 5-6   Level 7–8

1 Simplify the following ratios where possible.

a 6 : 8   **b** 10 : 2   **c** 11 : 33

d 6 : 7   **e** 12 : 8   **f** 24 : 18

2 **Write** two equivalent ratios for each of the following ratios.

a 1 : 2   **b** 4 : 9   **c** 3 : 6

d 1 : 6   **e** 4 : 1   **f** 20 : 30

3 At the Iowa State Fair, contestants can compete in a Pigtail, Ponytail, Braid, Mullet and Mohawk competition. Competitors vie for the Blue Ribbon Award in one of the four hairstyle contests.
A mullet is a hairstyle where the front and sides are short and the back is long. Called the 'Hunnic' look in the 6th century, mullets have been described as 'business in the front, party in the back'.

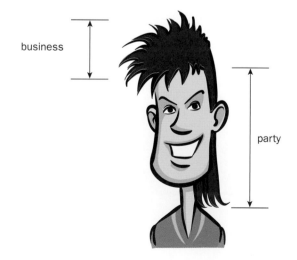

business

party

Some people suggest that the ratio of the length of the hair in back (the party) to the length of the hair in front (the business) would be a good measure of 'mulletness'.

A large ratio of party : business is desired.

a **Describe** what that means in terms of which mullets are likely to win.

b Determine who wins the mullet contest given the following ratios. Show your working.

4 Despite heavy competition, the work force is made up of 36% college graduates and 34% high-school graduates. The remaining are neither college graduates nor high-school graduates.

a Rewrite the percentage of college graduates, high-school graduates and neither as simplified fractions.

b Find the ratio of college graduates : high-school graduates : neither.

c Find the simplified ratio of college graduates : high-school graduates : neither.

5 a Is the relationship shown in the graph proportional? **Explain**.

b What is the constant of proportionality?

c Find the value of y when x is 24.

d Find the value of x when y is 108.

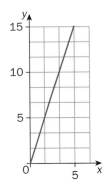

6 In 2016, the International Olympic Committee created the Refugee Olympic Team (ROT) to draw attention to the refugee crisis occurring worldwide and to honour those who 'have no home, no team, no flag, no national anthem'. In any other Olympics, the 10 athletes would have been on competing national teams, but this year they would cooperate as a unified team. The athletes and their country of origin are listed below.

Name	Country of origin
Rami Anis	Syria
Yiech Pur Biel	South Sudan
James Nyang Chiengjiek	South Sudan
Yonas Kinde	Ethiopia
Anjelina Nada Lohalith	South Sudan
Rose Nathike Lokonyen	South Sudan
Paulo Amotun Lokoro	South Sudan
Yolande Bukasa Mabika	Democratic Republic of the Congo
Yusra Mardini	Syria
Popole Misenga	Democratic Republic of the Congo

a **Write** the ratio of the number of team members from each country of origin to the total number of team members.

b Paulo Lokoro ran the 1500 m race in 4 minutes and 4 seconds, while Yiech Biel ran the 800 m race in 115 seconds. If they ran a 400 m race against each other, is it possible to tell who would win? Why or why not?

c As a refugee, Yusra Mardini had to swim and push the dinghy that she used to flee from Syria. At the Olympics she swam the 100 m freestyle event in 69 seconds. Anjelina Lohalith ran the 1500 m race in 4 minutes 47 seconds. Who had the faster speed? **Show** your working.

7 **Use** proportional reasoning to solve for the variable. Round to the nearest tenth where necessary.

a $2 : 9 = 10 : p$ **b** $36 : 12 = 54 : m$ **c** $11 : 33 = y : 11$

d $\dfrac{3}{t} = \dfrac{5}{11}$ **e** $\dfrac{8}{21} = \dfrac{f}{4}$ **f** $\dfrac{2.4}{5.7} = \dfrac{9.2}{k}$

8 Out of 2000 competitors at the World Iron Man triathlon, a total of 1815 completed the race.

   a Assuming that male and female competitors dropped out at the same rate, find the number of male and female finishers given the ratio of male competitors : female competitors is 20 : 13.

   b Find the simplified ratio of those who finished the race to those who did not.

9 A recent study found that the pay ratio of college graduates to non-college graduates over a lifetime is 184 : 100. On average, it is found that a person without a college degree will make $1.3 million ($1 300 000) in a lifetime. Given this ratio, how much would the average college graduate make in a lifetime?

10 If the ratio $x : y$ is 4 : 5 and the ratio $x : z$ is 2 : 3, then what is the ratio of $y : z$?

11 In California, there is a 'wildfire season', when wildfires are most common. These can be incredibly destructive events, with some burning as many as 1300 km². In 2007, the Moonlight Fire, caused by a lightning strike, affected people in three US states, requiring firefighters from across the United States to work together to get it under control. When the fire had burned 110 km², 1900 firefighters were battling the blaze. By the time it was under control, 2300 firefighters were putting out the fire that burned a total of 263 km². Because of this amazing cooperation, nobody perished and only two buildings were destroyed in the Moonlight Fire.

   a Represent the ratio of firefighters to area burned at each stage of the fire in three different ways. Show your working.

   b If the ratio of firefighters to area burned was maintained, how many firefighters should have been fighting the fire by the time it was under control? **Show** your working.

   c **Suggest** reasons why the number of firefighters was less than the amount you calculated in part **b**.

12 At the 1996 Olympics, Michael Johnson won the 200 m race with a time of 19.32 seconds. Donovan Bailey won the 100 m race with a time of 9.84 seconds. Both men claimed to be "The World's Fastest Man".

a Who had the faster speed? **Show** your working.

The media reported that Johnson was a faster runner than Bailey.

b Explain why this interpretation makes sense.

c Bailey said that this is not a fair comparison. **Explain** why he is justified in saying so.

d In order to see who the 'fastest man' was, the two ran a race of 150 m in 1997. Using their results from the Olympics, **predict** the time for each in the 150 m race.

e Bailey won the race in 14.99 seconds. How does this compare to his speed for the 100 m race? **Show** your working.

13 The Human Genome Project is an international collaboration to map the three billion base pairs in the human genome, the complex set of genetic instructions. Over 1000 scientists from six different countries collaborated on the project, which took over 13 years to complete. Genome size (measured in picograms) is the amount of DNA contained in a single genome. One picogram (pg) contains 978 million base pairs.

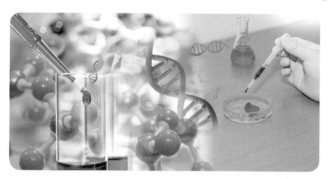

a Construct a table with the number of picograms in one column and the corresponding number of base pairs (in millions) in the other. Show at least four rows.

b Is this a proportional relationship? **Justify** your answer.

c Write an equation for the data in your table.

d Find the number of picograms if there are 11 247 million base pairs. Show your working using both a proportion and your equation.

Summative assessment

What is fair competition?

The Olympic Games, both summer and winter, are the world's leading sports competitions with more than 200 countries participating. All Olympic sports are organized by gender to insure fairness; men and women do not compete against each other. Some events, such as weightlifting and boxing, are further organized by weight, because of the unfair advantage larger athletes might have over smaller athletes. But what about height? Does the height of an athlete play a major role in the athlete's overall performance? Should some competitions also be organized by athletes' height?

Reflect and discuss 11 (as a class)

- Think of three sports in which taller athletes have a distinct advantage. Where do shorter athletes have an advantage?

- There is speculation that taller competitors have an advantage in short, fast-paced events. Do you think this is true? Give an example or two to justify your answer.

The Proportional Olympic Games

1 Track event

   a Your teacher will assign you a short (less than or equal to 400 m) individual track event to analyse.

   b Research who won gold, silver and bronze at the last summer Olympics in that event for both male and female competitors. In a table like the one here, record the name and height of each athlete as well as the time it took to run the race.

Medal	Female athlete	Height (m)	Time (s)	Male athlete	Height (m)	Time (s)
Gold						
Silver						
Bronze						

Suppose sprinters in this short race ran distances proportional to their height. Would that make a difference to who would have won the event?

c   Find the average speed of each runner (ratio of distance : time) and represent it as a decimal. Show all of your working.

d   Suppose the gold medallist runs the original distance of the event and the other medallists run a race that is proportional to their own height. Calculate the distance the silver and bronze medallists would have to run. Show all of your working.

e   Assuming runners run at a constant speed, how long would it take for the silver and bronze medallists to run their proportional race? Show all of your working.

f   Based on your results, who would have won the gold, silver and bronze medals in the proportional competition?

g   Are the new race times closer than they were before? Did this occur in both male and corresponding female events? Does height seem to matter more with women or with men?

h   Repeat the same process, assuming the silver medallist runs the original event and the other two competitors run an event that is proportional to their height.

i   Repeat the same process assuming the bronze medallist runs the original event and the other two competitors run an event that is proportional to their height. Organize all of your results in a table.

j   If you do make running distance proportional to height, could men and woman compete in the same event? Justify your choice.

k   Based on your results, is height an unfair advantage in the track competition you selected? Justify your response.

2 Research your own event

a What other individual event in either the summer or winter Olympics do you think might provide an unfair advantage to taller athletes?

b Repeat the same process you did with the track event to determine whether or not height seems to matter. Organize your results in a table and be sure to show all of your working.

c On the basis of your results, is height an unfair advantage in the event you selected?

d Given your height, select one event from the two you have studied and calculate the proportional distance you would have to run/skate/cycle etc. against the gold medalist. Does that seem achievable?

Reflect and discuss 12

- How precise do your calculations have to be? To how many decimal places should you round? Explain.

- Should athletes compete in height classes for select events in the Olympics? Justify why or why not.

- Do you think you can make the decision based on the information you have? If you were to do this analysis again, is there any other information you would include?

- What makes for fair and healthy competition? Explain.

- Which is more about being equal, competition or cooperation? Explain.

(2) Probability

Determining the odds of winning a game is an everyday example of the usefulness of probability. Would you make a decision if you were certain of the outcome? What if you were 55% positive? How certain would you need to be? The study of probability looks at the likelihood of events occurring, but these events can be much more serious than a game.

Globalization and sustainability

Commonality and diversity

With almost 8 billion people living on the planet, how much do any of us have in common? What's the likelihood of two people in a room of 30 having something simple in common, such as the same birthday? You might assume that it is highly unlikely. However, probability can be used to show that two people in that room will almost certainly share a birthday.

On a larger scale, citizens of different countries are often characterised by specific personality traits. How likely is it that a Canadian you meet is friendly or that a person from France is a good cook? Research suggests these stereotypes are more myth than fact, and that people are as diverse in terms of personality as they are physically. The study of probability could help you to separate fact from fiction.

Criminal justice

Criminal justice used to be in the hands of a single judge or maybe a jury, hearing testimony from witnesses and arguments from lawyers. With the advent of forensic science and DNA technology, more evidence can be provided to back up claims made in criminal trials.

With precise fibre and hair analysis it is much easier to show that someone was or wasn't at a crime scene. These analyses aren't infallible proof, but they determine the probability of an event occuring. When there is only a 0.00002% chance that someone was involved in a crime, can they be ruled out as a suspect?

In spite of the latest technology, there are still times when the criminal justice system fails. One famous case concluded that a male suspect was guilty based on DNA evidence that suggested he had been in contact with the victim. However, the quality of the DNA sample was questionable because the suspect had a skin condition which caused him to shed skin more frequently than most people. These factors undermined the case against him. Does DNA evidence offer conclusive proof or does it simply make for a more educated guess?

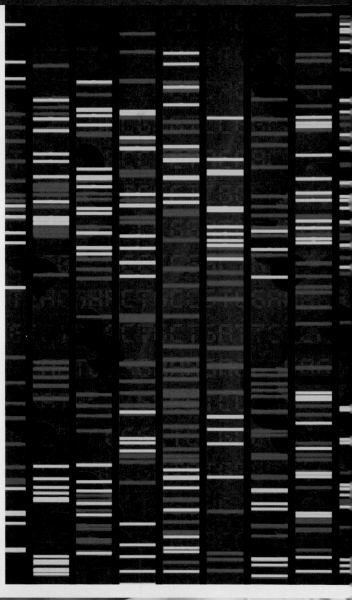

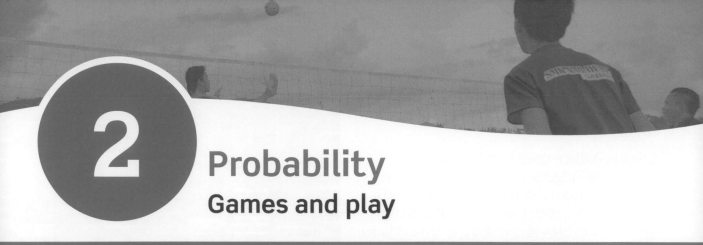

2 Probability
Games and play

KEY CONCEPT: LOGIC

Related concepts: Representation, Systems and Justification

Global context:

In this unit you will explore games that people from different cultures have created as part of your study of the global context **personal and cultural expression**. Games have been a primary means for cultures to promote social interaction. Despite the focus on leisure time and togetherness, many of these games also promote friendly competition and rivalry. Relying heavily on similar mathematical principles, it just might be that the person who knows how to analyse these games will be the one to win.

Statement of Inquiry:

A logical system of representation can help explore and analyse games that humans play.

Objectives

- Representing the likelihood of an event as fraction, decimal and percentage
- Modeling sample spaces in organized lists, tables and tree diagrams
- Calculating the theoretical probability of an event
- Designing and conducting simulations to calculate the experimental probability of an event

Inquiry questions

F
What makes something logical?
What does it mean when an event is 'likely to happen'?

C
How can logic be used with different representations?
How can you represent the likelihood of an event occurring?

D
Can winning be calculated or is it luck?

ATL1 Critical-thinking skills

Evaluate and manage risk

ATL2 Communication skills

Organize and depict information logically

You should already know how to:

- order numbers
- round numbers correctly
- represent numbers as fractions, decimals and percentages

Introducing probability

Every culture has games. The ancient Romans had tabula, a version of backgammon. The Mayans had Pitz, a ballgame played on courts that is still played in parts of Central America. Mancala is a count-and-capture game played all over Africa and the Caribbean. Medieval Europe had its chivalrous tournaments. Since then, we have invented more board games, social games, role-playing games and, most recently, video games. Everywhere you look, and in every time period, you will find people playing games.

Games and play have been social events since their first introduction. They help express cultural and religious views, and even help prepare players for the real world. Anyone from anywhere can learn the rules of a game and play it in their own setting. Similarly, mathematics is the only language that is shared by all humans, regardless of gender, religion or culture. It is no surprise, then, that many of the games we play rely on mathematical principles. Is it the game that crosses cultural boundaries or is it the underlying mathematics?

Reflect and discuss 1

Pairs

In pairs, think about a time that you've read or heard someone use one of the words below.

Chance(s)   Likelihood   Odds

- Write a sentence using one of these words that might describe the weather.
- Write a newspaper headline using one of these words related to a sport.
- Write a sentence that you might read about a game at an arcade or carnival using one of the words.
- Give an example of how knowing the likelihood of something happening increased your willingness to try.

Events and outcomes

There are many events—such as birthdays, holidays and vacations—during which people in various cultures play games. In mathematics, an *event* is something that involves one or more possibilities or *outcomes*. The games you play are probably full of these kinds of events, since there are often several possible outcomes during a player's turn.

Activity 1 – Matching events and outcomes

1 Select an event from the first column and match it with the possible outcomes in the last column.

	Events	Possible outcomes	
answering a question			2 Monday
			3 Spanish
			tails
flipping a coin			boy false 1
			Wednesday
			Sunday
rolling a six-sided die			5
			4 Japanese
			French
			heads
having a baby			6 Tuesday
			Thursday Saturday
choosing a day			true German
			Friday
selecting a language			girl
			Cantonese

2 Create a new row with your own example of an event from a game and list all of its possible outcomes.

If two or more events happen at the same time, they are called *compound events*. In a video game, picking from a variety of characters and then selecting a special power from several choices for that character is an example of a compound event.

Representing the sample space

The *sample space* of an event is the set of all possible outcomes. For example, 'semut, orang, gajah', or 'ant, person, and elephant', is a game played in Sumatra, very similar to 'rock, paper, scissors'. Players pump their fist on a count of three and then, on four, they show one of three outcomes: a pinky (semut), index finger (orang) or thumb (gajah). In this game, orang beats semut, gajah beats orang, and semut beats gajah (because the ant can crawl in the elephant's ear and drive it crazy). If the players choose the same outcome, it is considered a tie. Whoever wins 2 of 3 matches, wins.

It is very helpful to be able to organize all of the outcomes of an event like this one. One way to represent them is to use a list, while a second way is to draw a tree diagram.

 ATL2

Activity 2 – Representing the sample space using a list and a tree diagram

1   Write down a list of all of the possible combinations of outcomes for two players playing one round of 'semut, orang, gajah'. For example, 'semut-semut'.

2   How many different possibilities are there?

3   In how many of the possibilities does the first player win? In how many of the possibilities does the second player win?

4   In how many of the possibilities is it a tie?

A tree diagram represents possibilities in a more pictorial way. Different aspects of a game can be represented in a tree diagram.

▶ Continued on next page

5 Fill in the blanks with the possible outcomes for each of the players for one round of 'semut, orang, gajah'.

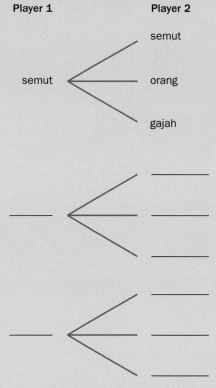

6 How many times would player 1 have won? How many times would player 2 have won?

7 How many times would it have been a tie?

8 Write down some similarities and differences between a list and a tree diagram.

9 A tree diagram can also be used to represent the possible outcomes for a single player for each of the three rounds. Complete a copy of the tree diagram that has been started below.

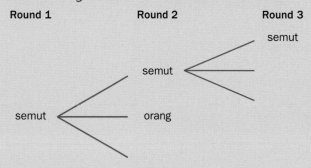

10 How useful is this tree diagram? Explain.

11 Which tree diagram representation do you prefer? Why?

Another way to represent a sample space is to use a table. Where a list and a tree diagram show all of the possibilities, a table can show all of the options or it can group outcomes together.

ATL2

Activity 3 – Representing the sample space using a table

1 a Copy the table below and fill in the rest of it to represent all of the possible outcomes for two players playing one round of semut, orang, gajah. Use the symbols 's', 'o', and 'g' to represent outcomes in the table.

		Player 2		
		semut (s)	orang (o)	gajah (g)
Player 1	semut (s)			
	orang (o)			
	gajah (g)			

   b Of the three representations (list, table or tree diagram), which do you prefer to use to represent all of the outcomes of the game? Explain.

   c Can you use a table to represent the outcomes of two rounds of semut, orang, gajah? Explain.

2 Suppose your soccer team is playing in a three-week competition and will play one game each week. The possible outcomes of each game are: a win, a loss or a tie.

   a Represent the sample space for the three-week competition with a list and a tree diagram.

   b To rank the teams, the organizers assign 10 points for a win, 5 points for a tie and 0 points for a loss. Given this information, what are the total points possible for each of the outcomes in your sample space? Add that information to both your list and tree diagram.

   c To qualify for the championship, your team must have at least 20 points accumulated and no more than one loss. How many outcomes are possible that would qualify a team for the championship? Use the table below to organize your sample space, given this criterion.

	Has no more than 1 loss	Has multiple losses
20 points or more		
Fewer than 20 points		

▶ Continued on next page

d Given the table organization, what information can you quickly find? List at least six facts.

e How many outcomes include multiple losses?

f How many outcomes have fewer than 20 points?

g Given the results in your table, do you think most teams would qualify for the championship? Explain your answer.

Reflect and discuss 2

- How do lists, tree diagrams and tables relate to one another?

- What limitations and/or strengths does each model have in representing a sample space?

Example 1

(Q) Before a cricket match, the umpires meet with the captains of the two teams to decide who will bat first. In general, the home team captain flips a coin and the visiting team captain chooses heads or tails. The winner of the coin toss picks one of two choices: to bat first or to field first.

a Represent the sample space (outcome of coin toss, outcome of their choice) using a list, a tree diagram and a table.

b How many different outcomes are possible?

c A team captain believes heads is luckier than tails. Justify whether you think he is correct or not.

(A) **a** The sample space is the list of all possible combinations of a coin toss, followed by the decision to bat first or field first.

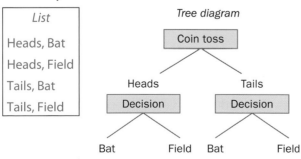

▶ Continued on next page

	Bat (B)	Field (F)
Heads (H)	HB	HF
Tails (T)	TB	TF

Note: the headers of a table can be created in many ways to organize your list or tree diagram. The way in which you title your headers supports the analysis you can do.

b There are 4 possible choice combinations.

You can count the combinations by examining your list, counting the number of branches at the bottom of your tree diagram or by counting the entries in your table.

c Heads and tails are equally likely. There is no lucky side.

Because half of the outcomes involve heads and half involve tails, both sides have an equal chance of being the winning side. This is also why they flip a coin, because both outcomes are equally likely!

Did you know?

A *fair coin* is one which, when flipped, is as likely to land on one side as the other. The chance of either side winning is truly 50%. However, most coins have one side that is slightly heavier than the other, which influences the results. It has also been shown that coins that are flipped vigorously tend to land more often on the side that was facing down before it was flipped. However these factors may affect the results by less than 1%, so they are a very minor issue.

'World Flip a Coin Day', where you are invited to make all decisions using the flip of a coin, is celebrated on June 1st every year.

Practice 1

1 List the possible outcomes for each of these events:

  a A fair coin is tossed.

  b A six-sided die is rolled.

  c Classes are chosen for a Year 2 student by a counselor.

  d A date is chosen in February.

2 On your birthday, your family has decided to let you choose how to celebrate it, with some conditions. They would like to take you out for dinner, and do an activity as a family. They provide you with some options (on the next page) and ask that you make a decision by tonight.

▶ Continued on next page

Family dinner at a restaurant	Activity
Italian	Laser tag
Japanese	Go-cart
Moroccan	Arcade

a Represent the sample space as a list and a tree diagram.

b Find the total number of celebration options (outcomes).

c Given this sample space, which celebration option would you pick? Explain.

d If you recommended another activity option, how would that affect your sample space? Explain.

e Suppose your family is on a strict budget. The family wants to celebrate your birthday, but also wants to make sure that they are mindful of their spending. Given the prices below, what are the total costs of each of your previous celebration options? Show these in both the list and the tree diagram.

Family dinner at a restaurant	Activity
Italian ($55)	Laser tag ($64)
Japanese ($68)	Go-cart ($59)
Moroccan ($62)	Arcade ($30)

f You have decided that there are two very important factors in your decision-making: the total cost of the celebration, and whether or not it includes going go-carting. Consider your overall sample space and fill in a copy of the table below, given the information you've gathered so far. Indicate the number of options in each of these categories.

	The option includes go-carting	The option does not include go-carting
Celebration cost is under $120		
Celebration cost is over $120		

g Find the number of outcomes that are under $120.

h Find the number of outcomes that include going go-carting.

i Find the number of outcomes that are under $120 and include going go-carting.

j Given the information in your table, what celebration would you choose? Justify your choice using evidence from your table.

▶ Continued on next page

3 *Clue* (also called *Cluedo*) is a board game in which players try to guess the identity of a criminal, the weapon that was used and the location of the crime. The possibilities for each are given in the table below.

Possible suspects	Potential weapon	Possible location
Mr Green	rope	library
Miss Scarlett	knife	study
Colonel Mustard	candlestick	dining room
Professor Plum	revolver	kitchen
Mrs White	lead pipe	ballroom
Mrs Peacock	wrench	conservatory
		billiard room
		lounge
		cellar

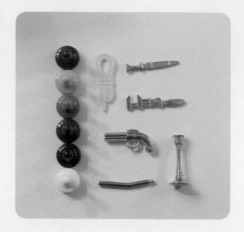

a Find the number of possible combinations. Explain how you arrived at your answer.

b The original game was produced in the United Kingdom in 1949 and has undergone several changes. Suppose a new suspect (Dr Orchid) and a new weapon (poison) are introduced. Write down the new sample space and find the number of possible combinations.

4 *Lu-Lu* is an old game played on the islands of Hawaii. There are four volcanic stone discs; one disc has 1 dot on it, one has 2 dots, one has 3 dots, and one has 4 dots. The reverse side of each disc has no dots on it. Players throw the discs in the air and score points for the total number of dots that appear face up.

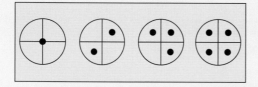

a What method would you choose to represent the sample space for all the possible outcomes? Justify your choice.

b How many different outcomes are there?

c A player earns points equal to the total number of dots she sees. Find all of the possible scores that a player can receive on one throw. In how many ways can each of the scores be earned?

The mathematics of probability

Representing probability numerically

The *probability* of an event occurring is a number that describes the likelihood that it might occur in the future. When stating the probability of an event occurring, the notation is *P*(event) = number. For example, P(rain tomorrow) = 75% indicates that there is a 75% chance (or probability) of rain tomorrow.

Activity 4 – Classifying events

The probability of an event occurring can be written as a fraction, a decimal or a percentage. For example, if it is Monday, then it is 100% likely that tomorrow will be Tuesday. You could also say that there is 0% chance that tomorrow will be Friday, no matter how much you wish otherwise.

1 With these examples in mind, where might 100% fall on the number line below? Where would you place 0% on the line? Copy the number line onto your paper and add those numbers in the appropriate spots.

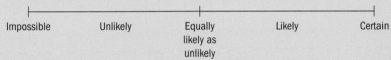

| Impossible | Unlikely | Equally likely as unlikely | Likely | Certain |

2 Use the sentences below to determine what numbers/values might correspond with Impossible, Unlikely, Equally likely as unlikely, Likely and Certain. Add these numbers/values on to your number line from step 1.

- 2 out of every 3 of people prefer chocolate over vanilla ice cream.

- There is a 50% chance of rain tomorrow.

- FC Barcelona's poor record has earned them the low odds of 10 to 1 of winning the championship.

- There is a 0% chance that you will draw a purple jellybean from a bag of only red and blue jellybeans.

- In 2013, sports analysts calculated that the probability of pitching a perfect game in baseball is 0.00022.

- If you put cards that spell out M, A, T, H, into a hat and pick one, then there is a $\frac{1}{4}$ probability that you will draw a vowel.

- Your teacher writes P(red) = $\frac{4}{5}$, and says, 'The probability of drawing a red marble is four fifths'.

▶ Continued on next page

3 Write down a sentence that you could add to the ones above that would fall between Unlikely and Equally likely as unlikely.

4 Analyse the numbers you've collected on the number line in this activity. What trends do you notice?

5 What range of numbers describes the likelihood of an event occurring? Write your answer in this format:

$$[S, L]$$

where S = the smallest possible value that is included, and L = the largest possible value that is included.

6 When you calculate the probability of an event occurring, the number calculated represents the likelihood of it happening in the future. Use conclusions drawn from the activity and your knowledge of a number line to complete the table as it relates to the number line beneath it.

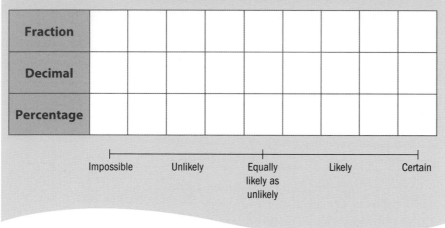

Fraction									
Decimal									
Percentage									

Impossible   Unlikely   Equally likely as unlikely   Likely   Certain

---

ATL1 ## Reflect and discuss 3

- Classify the following events as impossible, unlikely, equally likely as unlikely, likely or certain. Explain your reasoning.

  ○ The probability of getting a hole-in-one in golf is 1 in 12 500.

  ○ 1 in 14 000 000 is the chance of playing the lottery and winning.

  ○ The chance of someone with a university degree finding employment is 91%.

  ○ 1 in 3 attempts to climb Annapurna (in the Himalayas) results in injury or death.

- Despite your classification, are there probabilities in the questions above that would affect your behavior or decisions? Explain.

Practice 2

1 Complete the table below by classifying each event as impossible, unlikely, equally likely as unlikely, likely or certain, based on its probability. The first one has been done for you.

Probability statement	How likely is this event?	Make up a scenario. What decision might you make given this probability value?
$P(\text{rain}) = 0.23$	*Unlikely*	*'When I go outside, I probably won't need to bring an umbrella.'*
$P(\text{red marble}) = \dfrac{3}{4}$		
$P(\text{perfect score}) = 80\%$		
$P(\text{win}) = 0.5$		
$P(\text{girl}) = \dfrac{3}{31}$		

2 You and your friends are playing a game called 'Spin to Win' using the spinner to the right. Each sector of the spinner is the same size.

a Represent each value below as a decimal and a percentage, and then determine whether the event is impossible, unlikely, equally likely as unlikely, likely or certain.

(i) $P(\text{even number}) = \dfrac{4}{8}$ **(ii)** $P(\text{multiple of 4}) = \dfrac{2}{8}$

(iii) $P(\text{odd number less than 6}) = \dfrac{3}{8}$ **(iv)** $P(\text{less than or equal to 5}) = \dfrac{5}{8}$

(v) $P(7) = \dfrac{1}{8}$ **(vi)** $P(\text{positive number}) = \dfrac{8}{8}$

b You are creating the rules for your own Spin to Win game so that you are most likely to win. How might you use the probability likelihoods you determined in question **2** to strategize?

3 Classify the following as impossible, unlikely, equally likely as unlikely, likely or certain. Explain your reasoning.

a The likelihood of an adult owning a car is 67%.

b The chances of being from Andorra (next to Spain) and being literate are 100%.

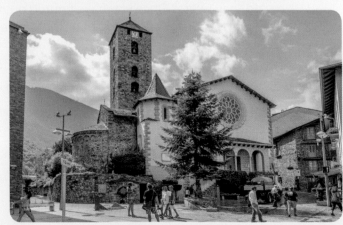

▶ Continued on next page

c The probability of a person living in Europe is 10%.

d There is a 16% chance of a magnitude 7.5 (or greater) earthquake in Bakersfield, CA in the next 30 years.

4 Choose a game that you like to play (video, board, sport, etc.) and write down an example of an event that fits each of the following classifications: impossible, unlikely, equally likely as unlikely, likely or certain. Assign a value to the probability and explain your choice.

Calculating probability

Up until now, probability values have been provided to you. Where do you suppose these values come from? How do you calculate the probability of events? In the following investigation, you will generalize the formula for calculating the probability of an event.

criterion **B**

Investigation 1 – Some simple probabilities

Spinners are often used in games, such as the one shown here.

The following probabilities can be calculated for the spinner.

$P(\text{even number}) = \dfrac{4}{8}$

$P(\text{multiple of 4}) = \dfrac{2}{8}$

$P(\text{odd number less than 6}) = \dfrac{3}{8}$

$P(\text{positive number}) = \dfrac{8}{8}$

1 Examine the spinner and the probability values. How do they relate? Once you have figured out their relationship, choose two examples and describe in logical steps how to arrive at that ratio.

2 How would you generalize the rule to calculate the probability of an event? Explain your reasoning.

Suppose instead of a spinner, you look at rolling one six-sided die.

The following probabilities can be calculated:

$P(\text{even}) = \dfrac{3}{6}$ $P(\text{prime}) = \dfrac{4}{6}$ $P(\text{multiple of 3}) = \dfrac{2}{6}$

▶ Continued on next page

3 Does the rule you generalized previously work here? Explain.

4 Use your rule to calculate P(number greater than 4) when rolling the die.

5 Write down a general rule for calculating the probability of any event. Make sure that your rule is able to translate to any event, not just a spinner or a die.

6 Verify that your rule works for three other examples of events (using a deck of cards, etc.).

Example 2

Q 'Red Rover' is a popular game that children play in the playground. Thought to have originated in the United Kingdom, the game is also enjoyed by children in Australia, Canada and the United States. Two teams form lines that face each other by joining hands. Then one team calls someone from the other team to see if they can break through their chain, by shouting 'Red Rover, Red Rover send _____ over!' The following opposing team members are still on the field:

Boys: Samuel   Oliver   Jermaine   Noah

Girls: Maryam   Emily   Jessica   Sofía   Chloe   Scarlett

Write down the sample space for the next name that could be called in the game and determine the following probabilities:

a P(girl)

b P(name ends in a vowel)

A The sample space is:

| The sample space is a list of all of the possibilities. |

Samuel   Oliver   Jermaine   Noah   Maryam Emily   Jessica   Sofía   Chloe   Scarlett

$$P(girl) = \frac{6}{10}$$
$$= \frac{3}{5}$$

$$P(girl) = \frac{number\ of\ girls}{total\ number\ of\ players}$$

Don't forget to simplify your fractions.

$$P(\text{name ends in a vowel}) = \frac{4}{10}$$
$$= \frac{2}{5}$$

Vowels are a, e, i, o and u.

$$P(\text{name ends in a vowel}) = \frac{number\ of\ names\ that\ end\ in\ a\ vowel}{total\ number\ of\ names}$$

Pairs

Activity 5 – Calculating simple probability

Your math class is going on a field trip! Your teacher has decided to let one of the students pick the destination. But, which student to choose? To be fair, your teacher has written each student's name on a token and placed all the tokens in a paper bag. Because there is exactly one token for each student in the class, this is an example of a uniform probability model. A **uniform probability model** means that all outcomes have an equal likelihood (chance) of occurring. In this case, it means that there are no repeat tokens.

Using probability, can you predict which student will be chosen?

1 With a partner, write down the names of all the students in your class neatly on a piece of paper. The students in your class are the sample space for this event as they represent all of the possible outcomes for this activity. This will be your data collection sheet for this activity.

2 How many people are in your sample space?

3 Calculate the probabilities of the following events. Write your answers as ratios. Then convert them to an equivalent percentage. Round to the nearest tenth, if needed.

- P(Girl)

- P(Boy)

- P(Name contains the letter 'A')

- P(Short hair)

- P(Wearing sandals)

- P(Wearing glasses)

- P(Wearing green)

- P(Girl and name contains the letter 'A')

- P(Boy and wearing green)

4 Brainstorm with your partner and come up with four probabilities that you'd like to find. Add them to the list of probabilities you calculated in step 3.

▶ Continued on next page

5 If only one name will be drawn from the bag, what do you predict will happen? Can you name a few characteristics of the person who is most likely to be chosen? Explain your reasoning.

6 Your teacher will simulate this drawing of a name. How did your prediction compare to the results? If there were discrepancies, what do you think could explain them?

Practice 3

1 You are playing an adventure video game that takes place on the Greek island of Crete. You are at a point where you can take a bus or a taxi to the small town of Kaliviani. From there, you can walk or cycle on a dirt road that will take you to the Balos Beach lagoon. Each possibility is equally likely.

  a Make a tree diagram to illustrate the sample space of transportation options.

  b Find the probability that you will take the bus and then walk. Show your working.

2 A friend is going to roll a six-sided die and says, 'If I roll an even number, we will play checkers. If I roll an odd number, we will play chess'. She states, 'I know we're going to play chess, because it's more likely I'll roll an odd number!' Is your friend correct? Which is more likely, rolling an odd number or an even number? Show your probability calculations as fractions and percentages. Use your calculations to confirm or refute your friend's claim.

3 The likelihood of being picked at random to go on a gameshow is 1 in 3 700 000.

  a What are three different ways to write this same probability value?

  b How likely is this event to happen? Classify the event as impossible, unlikely, equally likely as unlikely, likely or certain.

4 You've decided to adapt the ancient Greek 'shell game'. You use 8 shells and place them face down on the table. You randomly place a coin under one of the shells.

  a You ask a friend to pick one shell. Calculate:

    i P(finds coin)

    ii P(finds no coin)

  b Find the minimum number of coins you would need to hide if you wanted to increase the odds of picking a coin to over 60%, while keeping the same number of shells. Provide at least two examples, with evidence, that the chances are greater than 60%.

▶ Continued on next page

c In the original game, if you wanted to decrease the likelihood of P(finds coin) to less than 10%, how many shells would you need to add? Explain your reasoning with a diagram, probability calculations and an explanation.

5 The Roulette wheel was invented by the French physicist, inventor and mathematician Blaise Pascal in an attempt to create a perpetual motion machine in 1655. In this game, the wheel is spun and a white ball is set in motion around the dark brown track. Where the ball stops corresponds to a number on the outer ring, which players try to predict.

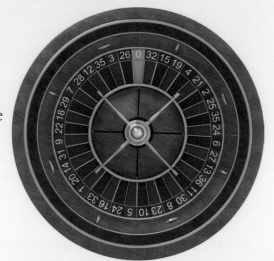

a Calculate:

  i P(4)

  ii P(red)

  iii P(odd number)

  iv P(4, 5, 6, 7, 8 or 9)

  v P(0)

b Originally Pascal's Roulette wheel had no '0'. In 1842, François and Lois Blanc added a zero for King Charles III of Monaco.

  i What would the P(22) be in 1655?

  ii What would the P(22) be in 1842?

  iii What effect does the addition of zero have on the game?

c Would you prefer to play roulette in 1655 or 1842? Explain.

ATL1   **d** Given the probabilities you calculated in the previous questions, is roulette a game that you think you have a good chance of winning? Explain.

6 A bag contains 4 red, 2 yellow, 1 white and 5 blue marbles.

  a If a marble is drawn at random, what are the chances that it will be blue or yellow?

  b If the marble is put back into the bag and another marble is drawn at random, what are the chances that it will be white?

  c If the white marble is *not* put back in the bag, what are the chances that the next marble drawn will be red?

▶ Continued on next page

7 The Monty Hall problem is a famous brain-teaser inspired by the popular game show, 'Let's Make a Deal' that first aired on US television in 1963. In the game, contestants are shown three doors (A, B and C), behind one of which is a fantastic prize, such as a car. Monty Hall, the game-show host, gives contestants the opportunity to select one door. If the contestant selects the door with the car behind it, they win it!

a Suppose the contestant selects door B. Monty then says, 'Now, doors A and C are right here and they remain closed. And door B is still closed and it *may* have a brand-new car behind it. Remember, the probability of your door having the car behind it is ____, and the probability of your door having nothing behind it is ____. Are you sure you want door B?' Write down the missing probabilities in Monty's response. Explain your reasoning.

b Once the contestant is satisfied with their choice, Monty says, 'Now, I know you want door B, but I'm going to do you a favor by opening one of the doors you didn't pick.' He then opens door C, and there is nothing behind it. Monty continues, 'Now, either door A or your door B has the car behind it, which means that probability of your door opening to the new car is ____.' Write down the missing probability in Monty's response. Explain your reasoning.

c It looks as though Monty Hall has changed the probability of winning in the middle of the game. What strategy would you use if you were in this situation? Would you stick with your original choice or would you change? Explain.

Complementary events

Sometimes there is a relationship between the outcomes of an event. One of these relationships is when outcomes are *complementary*.

Investigation 2 – Playing with probability

Dice of all shapes and sizes have been unearthed from ancient civilizations. 6-sided dice from 2000 BCE were found in Egyptian tombs and 4-sided dice were found with the oldest complete board game, the Royal Game of Ur, that dates back to Sumer in the 3rd millenium BCE. For this investigation, consider a dice with 12 sides to it, often used in role-playing games, with the numbers 1 to 12 on its sides.

Criterion **B**

1 Complete a copy of the table below. Represent probabilities as decimals rounded to the nearest hundredth.

▶ Continued on next page

	Column 1	Column 2
test A	P(rolling 12) =	P(rolling less than 12) =
test B	P(rolling an even number) =	P(rolling an odd number) =
test C	P(rolling 1, 2, 4, 5, 7, 8, 10, 11) =	P(rolling 3, 6, 9, 12) =
test D	P(rolling ≤ 3) =	P(rolling > 3) =

2 In each test, which is more likely: column 1 or column 2? Put a star next to the event that is more likely to occur.

3 What is the relationship between the events (column 1 and column 2) in each test? Explain.

4 What is the relationship between the probabilities of the events in each test? Explain.

5 Create a test E with two events involving a twelve-sided die that has this same relationship. Prove that the test E that you wrote follows the same pattern you identified.

6 Events that have the same relationship as the one you discovered between column 1 and column 2 are called *complementary events*. Write down a definition of 'complementary events' as well as a rule involving their probabilities.

7 Verify your rule with two examples that do not use a die.

8 Explain why your rule works.

Reflect and discuss 4

Suppose you have a set of cards, each with one letter on it. Your teacher lays the cards on the desk and they spell out:

- Using these cards, give an example of two complementary events. Show that they are complementary using their probabilities.

- What event is missing? P(vowel) + P(_____) = 1

- Can the probability of complementary events ever add up to a number greater than 1? Explain.

Practice 4

1 *Wheel of Fortune* is a television game show in which contestants compete to solve word puzzles to win cash and other prizes. It has aired on television in many countries, with different versions and hosts, and is still on air in the United States. One version of the wheel is shown on the right.

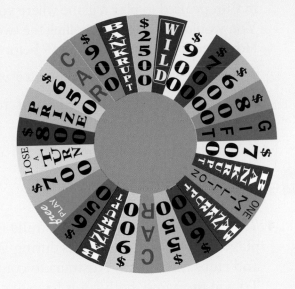

a If each sector is the same size, find the probability of landing on:

   i 'Lose a turn'

   ii a cash value of $700 or less

   iii 'Bankrupt'

   iv 'Car', 'Gift', 'Free play' or 'Wild'

   v a cash value greater than $800

   vi any blue sector.

b Give an example of two complementary events on the wheel. Demonstrate they are complementary using their probabilities.

ATL1

c If you knew the solution to the word puzzle, how would you decide whether or not to solve it or to keep spinning in the hopes of winning more money? **Explain**.

2 You are playing a game that involves throwing two six-sided dice at the same time and adding the numbers that are face up. For example, if you roll a 3 and a 2, the sum is 5.

ATL2

a Determine the sample space for this experiment. Represent the sample space as a tree diagram and as a table. Which representation is more effective? Explain.

b If you had to guess which sum would be rolled next, what would your choice be? Explain.

c Determine the probability of each of these events:

   i an odd number appearing next

   iii a number other than 7 appearing next

   ii a prime number appearing next

   iv a 7 or 11 appearing next.

d Name the complement of each event in part **c** and determine its probability.

▶ Continued on next page

3 Games with dice have been around since before the 12th century. For example, people in Europe used to play a game where you rolled two six-sided dice and found their sum. You would then continue rolling the two dice until you obtained that original sum again. However, if you rolled a sum of 7 before you rolled your original sum, you lost the game.

  a Explain why you think rolling a 7 was selected as the sum where you lose.

  b Is 'rolling a 7' the complement of 'rolling the same number again'? Explain your answer using probabilities.

  c Which sums would you prefer to roll on your first try? Explain.

ATL1 **d** This is actually a popular game in casinos. Would you want to play this game? Justify your reasoning.

4 *Two-up* is a game played in Australia in which two coins are tossed. *Two-up* was played by Australian soldiers during World War I and is played on Anzac Day in bars and clubs throughout Australia to commemorate the soldiers who fought in the war. To win the game, both coins have to come up heads.

  a Represent the sample space as a list, tree diagram or table.

  b Find the probability of getting two heads.

  c Find the probability of getting a head and a tail.

  d Find the probability of tossing two tails by using its complement.

5 *Tombola* is an Italian game that families would play in which the prizes were often symbolic. As in *bingo*, each player has a card like the one on the right.

Each card has three rows of nine columns, and each row has numbers in five of those columns. The announcer, called the tombolone, calls out numbers from 1 to 90 that are selected at random, announcing each with a little rhyme. A player wins when all of the numbers on their card are called.

11		36			65	71	89
7			44	52		76	84
1	26	37			62	70	

  a Calculate the following probabilities for the tombolone:

    P(odd)   P(number > 40)   P(multiple of 5)   P(prime number)

    P(repeated digit)   P(one of your numbers is called)

    P(one of your numbers is *not* called)

  b Which two events from part (a) are complementary? Show that they are complementary using probabilities.

  c Write down two other pairs of events in which each pair represents a set of complementary events. Show that the events are complementary using probability.

Formative assessment 1

criterion
C, D

The game show 'The Price is Right' started in the United States in 1956 and has been recreated in over 40 countries around the world. Contestants compete to win cash and prizes by playing a variety of games, often involving guessing the price of a product. During celebrity week (which raises money for many charities) former professional basketball player and sports commentator Charles Barkley played alongside contestants to offer advice during the games.

In one game, the contestant rolls five six-sided dice in attempts to win a car. Each die is labelled with a car on three sides and cash values on the other three sides ($500, $1000 and $1500). In order to win the car, the contestant must get a car picture on each die in a limited number of rolls. The contestant can win up to three rolls to get all five car pictures, and until the final roll can choose to take the cash values he/she rolls instead of trying to roll again for the car.

In a particular game, the contestant won the right to roll the dice twice.

1 Represent the sample space for the first roll of five dice using a list, tree diagram and a table.

2 On any one die, determine these probabilities:

  a P(car)

  b P($1000)

  c P(> $500)

3 Explain whether your results in step **2** make sense.

4 What is the likelihood of walking away with nothing (no car, no money)? Explain.

On the first roll, the contestant rolled two car pictures and the other three dice totalled $4000. The two car picture dice were removed and the contestant could choose to either take the $4000 cash or roll the three remaining dice one last time to try to win the car. If he rolled all three car pictures, he would win the car. However, if he rolled fewer than three cars, he would win whatever money was showing on the three remaining dice.

5 Represent the sample space for possible outcomes of rolling the three dice using a list and a tree diagram.

6 Charles Barkley advised the contestant to stop rolling and take the $4000. Use probability to explain why he would suggest that.

7 What would you suggest the contestant do? Justify your answer using probability.

If you want to see what happened, visit **priceisright.com**, and go to 'games, let em roll'.

Types of probability

The probability that you calculate is known as the *theoretical probability* because it represents what is *supposed* to happen, in theory. However, what happens in real life is not always the same as what theory predicts. This 'real' probability is known as the *experimental probability* of an event.

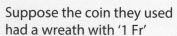

Investigation 3 – Experimental probability

Orville Wright was the first person to fly a plane, but did you know that it was partly thanks to Orville losing a coin toss? He and his brother, Wilbur, tossed a coin to see who would try to fly first, and Wilbur won. However, he stalled the plane and landed in the sand before he could actually take flight. Orville then got his turn a few days later and was successful!

Suppose the coin they used had a wreath with '1 Fr' (1 Swiss franc) in the center on one side and on the other side of the coin the image of the goddess Helvetia standing and holding a shield. Wilbur chooses the goddess Helvetia and Orville chooses the wreath. Whichever side is facing up will determine who flies first. While the brothers only flipped the coin once, does the number of times the coin is flipped change the outcome?

1   Suppose they decide to flip the coin once. What is the probability of the coin landing with the wreath side up? What is the probability of the coin landing with the goddess Helvetia side up?

2   Given those probability values, what is the likelihood of each occurring? Given this information, which side would you strategically pick to insure you fly first?

3   You and a partner are going to test this theory by flipping a coin 100 times to model this situation. To do this simulation you can flip an actual 1 Swiss franc or any coin and assign corresponding sides to the wreath and the goddess Helvetia.

4   Create a data sheet to record multiple trials. Keep in mind that you'll be doing 100 trials, so neatness and organization of your data will be important.

You can also simulate a coin toss using technology by going to **www.random.org** and using their Coin Flipper simulator. Be sure to choose to flip '1' coin and scroll through the different coins to find the Swiss 1 Franc.

▶ Continued on next page

5 Flip your coin 25 times and record the outcome each time. When done, calculate the experimental probability of obtaining a wreath and the probability of obtaining the goddess Helvetia.

6 Continue flipping your coin another 25 times and record the outcome each time. Again, calculate the experimental probability of each outcome.

7 To finish your investigation, flip the coin 50 more times. Record the outcome each time. For the last time, calculate the experimental probability of each outcome.

8 Copy the table below and organize your data collected from the simulation for the analysis of this investigation. Write your probability ratios as percentages. Round to the nearest hundredth, if needed.

	Theoretical Probability	Experimental probability after 25 trials	Experimental probability after 50 trials	Experimental probability after 100 trials
P(wreath)				
P(goddess Helvetia)				

9 Compare the values from your table to the theoretical probability of each outcome. How are they similar or different? If they are not the same, how close is the experimental probability to the theoretical?

10 What happens to the experimental probability as the number of trials increases? Explain.

11 Which is more fair: flipping a coin once or flipping it 100 times to determine who gets to fly first? Explain.

Reflect and discuss 5

- Why do the predicted probability values differ from experimental values? Explain.

- Are theoretical and experimental probabilities ever the same? Explain.

Earlier in this unit you considered the Monty Hall problem, where the probability of winning seems to change as the game is played. The question still remains: Is it better to stick with the original choice or is it a better strategy to change? This is a great place for a simulation, an activity that imitates a real-life event, to see what happens to the experimental probability as the game is played many times.

Activity 6 – The Monty Hall problem revisited

With a partner, simulate the Monty Hall problem by playing it at least 25 times each. Set up three 'doors' (e.g. cards or pieces of paper) and put a 'prize' behind one of them (coin, pen, etc.). One of you plays the role of Monty Hall and the other is the contestant. After the initial selection of a door, Monty should reveal what is behind one of the other doors where there is no prize.

1 One of you should always change your selection after the door is revealed and one of you should always stay with your original selection. Record the outcomes of each game in a table.

2 With your own data, calculate the probability of winning, using each strategy.

3 Combine your data with another pair and calculate the experimental probability of winning using each strategy.

4 As a class, combine your data and calculate the experimental probability of winning using each strategy.

You can play an online version of the Monty Hall problem if you search for Monty Hall on **mathwarehouse.com.**

Reflect and discuss 6

- Based on your simulation results, what strategy is the most effective for the contestant: keeping the original choice, or switching?

- If Monty Hall had *not* opened one of the doors during the game, what is the theoretical probability of P(win)? Did Monty Hall help the contestant's odds by opening a door? Explain.

- If you know P(win) with your initial choice, what is P(not win)? How does this relate to the theoretical probability of winning by switching your initial choice? Explain your answer using complementary events.

Practice 5

1 Represent the sample space for each:

a rolling a four-sided die

b days of the work week

c flipping two fair coins

d pathways from start to rooms A or B (see diagram).

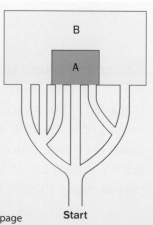

▶ Continued on next page

2 Based on the sample spaces in question **1**, find the theoretical probability of each of the following:

    a rolling a 2 on a four-sided die

    b taking a class on Tuesday or Wednesday during the work week

    c a heads and tails combination showing when flipping two coins

    d ending in room A (see diagram in question **1**).

3 A fair coin is tossed 72 times. A head comes up 12 times.

    a Find the experimental probability of tossing a fair coin and having it land on heads.

    b Find the experimental probability of tossing a fair coin and having it land on tails.

    c Show that the events in **a** and **b** are complementary using probability.

4 A six-sided die is rolled 30 times. If the die lands with a 1 or 6 facing up, the player wins. If the die lands on 2, 3, 4 or 5, the player loses. Consider the table of collected data.

Number rolled	Frequency
1	II
2	IIII
3	ℕℕ I
4	III
5	ℕℕ II
6	ℕℕ III

    a Find the experimental probability of winning and losing.

    b What is the theoretical probability of winning? What is the theoretical probability of losing?

    c Are you surprised by the results? Explain.

5 A spinner with three equal sections was spun 100 times. The results are as shown:

score	1	2	3
frequency	40	35	25

    a Find the experimental probability of spinning a 3. Compare the value to the theoretical probability.

▶ Continued on next page

b Find the theoretical and experimental probability of spinning a 1. Represent the experimental probability as a decimal.

c What would you expect to happen to the experimental probability of spinning a 1 if you were to spin the spinner 1000 times or more? Explain.

6 An 8-sided die is rolled, and a coin is tossed. Make a tree diagram of the sample space to find the probability of getting a head and an odd number.

7 Give an example of a method to simulate the outcomes of choosing one pair of jeans from a group of four pairs which are equally popular.

Activity 7 – Probability Fair

1 "Search for "probability fair" on the **MrNussbaum.com** website. There are five games in the fair and you need to win tickets before entering the fair.

2 Calculate the probability of each outcome listed on the wheel when getting tickets. Select the one with the highest probability. Did that sector win?

3 For each of the 5 games, answer the following questions:

a How many tickets is it to play?

b What is the probability of each possible outcome?

Play the game 5 times.

c Did the most likely outcome occur? Explain any differences.

d Given the payout for winning and the number of tickets it is to play, do you think you should play the game? Justify why or why not.

Simulations

Sometimes it just is not practical to perform an event several times in real life. There are other times when the theoretical probability of an event is not easy to determine. In both of these cases you could perform a *simulation*. A simulation imitates a real-life event and replaces the actual event with something that is much easier to perform, usually with dice or cards or a spinner. For example, if you wanted to simulate taking a multiple-choice test with four answers to each question, you could pretend to answer each question by using a spinner like the one on the next page. You would then decide which color represents getting the correct solution, since the theoretical probability of guessing the right answer is $\frac{1}{4}$.

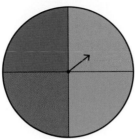

You could also roll a four-sided die and decide which number represents selecting the correct answer. You could even roll an eight-sided die and designate *two* numbers to represent selecting a correct answer.

The key is to find an event that you *can* do that has the same probability as the real-life event.

Activity 8 – Matching probabilities

1 Match each real-life event in the left column with an event in the right column that could be used to simulate it. Explain why they are a match.

Real-life event	Simulated event
Having a baby	Rolling a six-sided die
Being born on a specific day of the week	Spinning a spinner divided into 3 equal sectors
Making a shot when you score 75% of the time	Spinning a spinner divided into 4 equal sectors
Winning a gold, silver or bronze medal	Picking a card from a stack of cards numbered 1 through 10.
Getting accepted by one of six universities	Flipping a coin
10% of donors have type B blood	Rolling a 10-sided die
70% of DP students earn the IB Diploma	Picking one of 7 different marbles from a bag

2 Angel plays soccer and normally scores on half of his shots on goal. How could you simulate him taking six shots on goal in a game? Explain.

▶ Continued on next page

3 You are going to take a multiple-choice test with five questions, each with four possible answer choices. How could you simulate this? Explain your method.

4 Toy manufacturers are making miniature toys that are sealed in packaging. You don't know which toy you are getting until after you have purchased one and opened the package. If there are twelve toys in the series, how could you simulate finding out how many you would need to buy to collect all twelve toys?

Now that you have thought about how to simulate real-life events with easy-to-access materials, it is time to try designing a simulation and calculate experimental probabilities.

Activity 9 – Designing a simulation

Native Americans used to play a game called *The Bowl Game*, where they would toss four objects in the air (beans, nuts, buttons made of elk horns, etc.) and score points if they landed right side up or upside down in a woven basket. Points were scored in the following way:

> All four up = 5 pts
> 3 up, 1 down = 2 pts
> 2 up, 2 down = 1 pt
> 3 down, 1 up = 2 pts
> All four down = 5 pts

Suppose you were going to play this game five times.

1 Represent the sample space for tossing one object. What is the probability that it lands right side up or upside down? Explain.

2 Describe how you could simulate this probability using:

  a a die

  b a spinner

  c a third method of your choice.

3 Design your simulation so that you pretend to play five rounds, each requiring you to toss four objects.

4 With a partner, conduct the simulation at least 20 times and record your results in a table.

5 What is the experimental probability that your score will be more than 10 points? Show your calculations.

6 Combine your results with those of another pair of students. Recalculate the experimental probability.

▶ Continued on next page

7 Combine your results with those of another foursome. Recalculate the experimental probability.

8 Combine your results with those of the rest of the class. Recalculate the experimental probability.

9 If you repeated the simulation 100 more times, what experimental probability would you expect to find? Explain.

Practice 6

1 *The Game of Life* was a popular board game in the United States, first played in 1860 as *The Checkered Game of Life*. In the modern version, created in 1960, players move around the board and face events and decisions similar to those people encounter in life, including going to college, buying a car and having a family. Design and carry out a simulation to determine the experimental probability of all three children in a family being girls. Be sure to run your simulation at least 20 times.

2 *Ringball* is a traditional South African game that is related to both basketball and netball. Play takes place on a field (indoors or outdoors) where players pass a ball from one teammate to another and then, hopefully, throw the ball through a hoop on a goalpost. Suppose a player makes 75% of her shots when she tries to score. What is the experimental probability that she will score on three consecutive shots?

Design and carry out a simulation to answer this question. Be sure to run your simulation at least 30 times.

3 Create a simulation of the game from Formative assessment 1.

   a Write down the procedure for the simulation, including details on the experiment used to replace the real-life event.

   b Run the simulation 20 times. Summarize your results in a table.

   c How many times did you win a car after one roll?

   d Collate all the results of the class. What was the experimental probability of winning the car after one roll?

   e Calculate the experimental probability of winning the car in this scenario.

   f Would you have given the same advice as Charles Barkley? Justify your answer.

   g Calculate the theoretical probability of winning the car in this scenario. Are the experimental and theoretical probabilities different? Explain.

   h What advice would you give the contestant? Justify your answer.

▶ Continued on next page

4 In the game of *Skunk*, two players get five rounds to earn the most points. During a turn, the player rolls a six-sided die. If the player rolls a 1, the player gets 0 points and the turn is over. If the player rolls a 2, 3, 4, 5 or 6, then the value is recorded as that many points. If the player earned points, there is a decision to make: the player can take a risk and roll again, or keep the value earned. If the player rolls again, the value is added to the recorded score. If the player rolls a 1, the accumulated score goes down to zero.

a Write down the sample space for all of the possible outcomes of four rolls of the die.

b Calculate $P(1)$ on the first try.

c Find $P(> 1)$ on the first try.

d Find $P(>1, \text{then} >1)$.

e Find $P(>1, >1, \text{then} >1)$.

f Find $P(>1, >1, >1, \text{then} 1)$.

g Based on your calculations, is it wise for a player to continue risking to roll again? Use your calculations in your explanation.

Suppose you played *Skunk* with your friend Robin, and the first and second round results were those shown in these two tables.

round 1	Rolled	Score
You	2, 4, 5	11
Robin	3, 1	0

round 2	Rolled	Score
You	6, 5, 2, 4, 1	0
Robin	6, 6	12

h Examining round 1, what is the theoretical probability of rolling 2, then 4 and then 5? Were you lucky? Was Robin unlucky to roll 3 and then 1? Explain.

i Examining round 2, what is the theoretical probability of rolling 6, 5, 2, 4 and then 1? Should you have stopped after your fourth roll? Was it too risky?

j Before starting round 3, you have accumulated 11 points and Robin has 12 points. On your first roll in this round you've rolled a 2. Should you stop rolling? Is it worth the risk to roll again to widen your new lead?

Pairs

Formative assessment 2

You and a partner are to conduct a simulation of a real-life situation.

1 Determine the question that you want to have answered using a simulation.

2 Determine the sample space for your simulation.

3 Specify the procedure followed to conduct the simulation.

4 Create a spinner to simulate the outcomes – explain why the sectors of the spinner are the size they are.

5 If you were to conduct the simulation in another way, what other device could you use? Explain why that would also work in your simulation.

6 Conduct the simulation. Record the data in a table.

7 Complete the necessary calculations to answer your original question – show all steps.

8 Explain the degree of accuracy of your result(s).

9 Describe whether or not your result(s) make sense in the context of the problem.

Reflection:

Summarize your simulation in the context of the real-life situation it was representing.

What were the maximum, minimum and range of your data points?

Were the results of the simulation what you expected? Explain.

Do you feel you conducted the simulation enough times? How do you think the results would have changed if you did more trials?

If you were to conduct a simulation again, what would you do to improve upon the process you followed in this simulation?

Unit summary

The probability of an event occurring is a number between 0 and 1 that describes the likelihood that it might occur in the future. Words that describe the odds of an event occurring include: impossible, unlikely, equally unlikely as likely, likely and certain.

Sample spaces can be described using lists, tables and tree diagrams. For example:

List

rock-rock, rock-paper, rock-scissors,
paper-rock, paper-paper, paper-scissors,
scissors-rock, scissors-paper, scissors-scissors

Table

	rock	paper	scissors
rock			
paper			
scissors			

Tree diagram

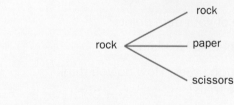

Probability is measured by the ratio of favorable cases to the total number of outcomes possible, or sample space. The ratio can be written as a decimal or percentage.

$$P(\text{event}) = \frac{number\ of\ favorable\ cases}{total\ number\ of\ outcomes}$$

Complementary events together make up the entire sample space and the sum of their probabilities is always 1.

Theoretical probability is what is expected to happen, based on theory.

Experimental probability is the ratio of actual occurrences of favorable cases to total trials conducted.

Experimental probability can yield different results from theoretical models, but the experimental probability gets closer to the theoretical value the more the experiment is performed.

Simulations can be designed to mimic real-life situations and can provide data that can be used to calculate theoretical probability.

Unit review

criterion **A**

⎧ 🖺 **Launch additional digital resources for this chapter** ⎫

Key to Unit review question levels:

Level 1–2 **Level 3–4** **Level 5–6** **Level 7–8**

ATL2 ❶ Make another sample space representation for:

a a list of outcomes for spinning a spinner and flipping a coin

b a table for rolling two dice

	1	2	3	4	5	6
1						
2						
3						
4						
5						
6						

c a tree diagram for three rounds of drawing a blue or green marble from a bag.

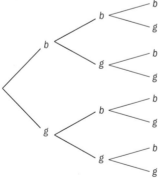

❷ *Who Wants to be a Millionaire?* is a British television game show in which players answer multiple choice questions and could win up to one million British pounds. Over 100 different international variations of the show have appeared since the show first began in 1998.

a Each question is multiple choice with four possible answers (A, B, C, D). What is the probability of guessing an answer correctly?

b Play stops as soon as a question is answered incorrectly. Represent the sample space for the outcomes of up to three questions answered in a row using a tree diagram and a list.

c If a player doesn't know the answer to a question, he can use one of three 'lifelines'. One of those is called 50/50, where two of the wrong answers are eliminated. What is P(correct answer) after the 50/50 lifeline option is used?

3 *Dungeons and Dragons* is a very popular role-playing game where players are given a character and embark on adventures in a fantasy land. A Dungeon Master controls the game, creating scenarios and challenges for the players. Many of the actions that are taken in *Dungeons and Dragons* rely on rolling dice that have 4, 6, 8, 10, 12 or 20 sides. Suppose you are rolling an 8-sided die.

a Calculate:

   i P(2)

   ii P(2, 3 or 4)

   iii P(even)

   iv P(<5)

   v P(9)

b Show that P(1, 3, 4, 7, 8) + P(2, 5, 6) = 1

c Provide a different example of complementary events and provide calculations to justify your example.

d Given probability values you've calculated in parts **a** and **b**, determine which events are impossible, unlikely, equally likely as unlikely, likely and certain.

4 The game *Miljoenenjacht*, which originally began in the Netherlands, is more widely known as *Deal or No Deal*. In the game, there are 26 boxes, each with a different amount of money in it. At the beginning of the game, the player chooses one box which he/she hopes has the highest prize, €5 million. During the rest of the game, the player chooses boxes from those that remain. With each selection, the player knocks that box and the amount inside of it off the game board. The amounts in the boxes are shown in the table at the top of the next page.

€ 0,01	€ 10.000
€ 0,20	€ 25.000
€ 0,50	€ 50.000
€ 1	€ 75.000
€ 5	€ 100.000
€ 10	€ 200.000
€ 20	€ 300.000
€ 50	€ 400.000
€ 100	€ 500.000
€ 500	€ 750.000
€ 1.000	€ 1.000.000
€ 2.500	€ 2.500.000
€ 5.000	€ 5.000.000

a With a contestant's initial selection of a box, find the following probabilities:

  i P(€5 million)

  ii P(amount > €100 000)

  iii P(amount < €75 000)

  iv P(at least €1 million)

  v P(an amount in the thousands)

  vi P(€0)

b After a few rounds, only the boxes below remain.

€ 0,01	
	€ 25.000
€ 0,50	€ 50.000
€ 1	€ 75.000
	€ 100.000
€ 10	€ 200.000
€ 20	
	€ 400.000
€ 100	
€ 500	€ 750.000
	€ 5.000.000

Find the following probabilities related to the next box selected:

  i P(the €5 million is left on the board)

  ii P(the €5 million box is opened)

  iii P(the box < €20)

  iv P(the box > €100 000)

c After the few rounds, is the contestant in a better position to win the 5 million? Justify your response.

5 In the game of *Risk*, players set up armies on a map of the world that acts as the game board.

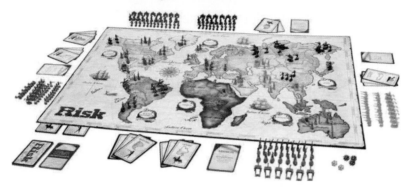

Players can add to their armies or wage war against armies in regions that are beside them. In these wars, each player rolls one or more six-sided dice, with each die representing a battle. With each battle, if the person initiating the war (the attacker) has a higher number, then he/she wins the battle. If the defender rolls a number that is equal to or greater than the attacker, then the defender wins the battle.

The number of dice rolled depends on the number of armies you have. When more than one die is rolled, whoever wins the most battles, wins the war. If each player wins the same number of battles, the defender wins the war.

a Suppose each player rolls just one die. Find the probability that the defender wins the war if the attacker rolls a:

   i 3   **ii** 1   **iii** 5   **iv** 6

b Suppose each player rolls two dice. Each player's highest number is compared with their opponent's, with the winner of each battle being the higher number for that battle. The lower numbers are then compared in the same way. Write the sample space for all possible outcomes of the defender's roll when the attacker has rolled a 2 and a 5. Create a table like the one started below, for example. And remember, if each player wins one battle then the defender wins the war.

Attacker rolls	Defender rolls	Battle 1 won by	Battle 2 won by	War won by
2, 5	1, 1	Attacker	Attacker	Attacker
	1, 2	Attacker	Attacker	Attacker
	1, 3	Attacker	Attacker	Attacker
	1, 4	Attacker	Attacker	Attacker
	1, 5	Defender	Attacker	Defender

c Find the probability of the defender winning the war (by winning one or both battles)

d What attacker rolls will result in the defender's probability of winning both battles be 1?

6 In the UK quiz show *Who Wants to be a Millionaire?* contestants continue to answer questions until they get one wrong.

The value of each question is shown here. If a contestant answers a question correctly, she moves on to the next question. If she answers incorrectly, she falls back to the amount in yellow. For example, if she is on question 9 and answers it incorrectly, she wins only £1000 and the game is over. If she answers any of the first four questions incorrectly, she wins nothing. A player can stop answering at any time and take the money she has earned.

15	£1 Million
14	£500,000
13	£250,000
12	£125,000
11	£64,000
10	£32,000
9	£16,000
8	£8,000
7	£4,000
6	£2,000
5	£1,000
4	£500
3	£300
2	£200
1	£100

a Knowing that each question has four answer choices, and assuming a contestant does not use a single lifeline, design a simulation to determine the average winnings a player can expect if she just guesses each answer.

b Record the data from your simulation in a table, showing the amount of money won each time. Average these over 20 trials.

c What is the experimental probability of P(> £16 000)?

d What is the experimental probability of P(£1 000 000)? Do you think anyone ever won that much money? If so, how?

ATL1 **e** Suppose you have used all of your lifelines and both the audience and the friend you phoned think the answer to a question is 'c'. You aren't sure they are correct. What is the highest dollar level where you would choose answer 'c' instead of stopping and taking the money you have earned? **Justify** your choice.

Summative assessment

It's time to relax and enjoy some game time with your fellow classmates!

Your assignment: You will be having a 'Games Day' in one or more of your upcoming classes, where all of the games that are created and played will be created by the students of your class.

Part A: Develop a game as partners

You and your partner will be creating a game for Games Day. The game you create should be novel and be quick to play (no more than 5 minutes). You can use any media to create your game, including (but not limited to) the use of technology and/or dice, cards, a spinner and coins. Think about what kinds of games are fun to play and how you can incorporate those elements into your game.

Make a To Do list:

a Create your game. Be thoughtful about the requirements that your game should be unique and exciting, be quick to play and have simple and compound events. It must be a fair game, not biased toward any particular player.

b Write a report for your teacher that outlines important characteristics about your game. Your report must include:

- instructions on how to play your game and the goal of the game

- a paragraph explaining the strategy for how to win the game

- two representations of the sample space of your game (list, tree diagram or table)

- the theoretical probability values you have calculated associated with your game, both simple and compound theoretical probabilities

- explanation of the mathematical reasoning associated with the construction of your game (for example, the smallest section on a spinner could have the highest reward), taking into account individual theoretical probabilities

- an activity sheet to log actual events that occurred when others played your game, which should include a table that allows you to quickly list the game player's name, a short summary of what happened during the game (Did they roll a 6? Did they draw a card with an 'H'?, in what order?, etc.) and the outcome (who won the game etc.); this information will be used to calculate the experimental probability values associated with your game being played, so make sure you set up an appropriate activity sheet to record all relevant data.

Part B: Games Day (as partners)

Pairs Your teacher will divide the class into groups, either first shift or second shift. For half the time you will be playing classmates' games and for the other half, some of your classmates will be playing your game.

The objective is for your team to have fun and socialize while playing friendly games against each other, making sure you collect valuable data on different games (including your own), so that you can analyse it in a summary analysis report. Take both objectives into account as you take part in Games Day. Your roles as game hosts and as game players are equally important.

During your game hosting:

Your team will set up your game and wait for classmates to come and play it.

Make sure you have:

- any important materials needed for the game, including clear instructions on how to play the game

- an activity sheet (in your report) to record all necessary data.

Work with your partner to make sure you are both participating in hosting the game and recording important data in your activity sheet. You will need to inform players about your game, including giving clear directions on how to play and the goal of the game. Don't forget to record events on your activity sheet.

During game play time:

You and your partner must get to know the games being offered and play as many different games as you can in the time allotted. Decide which games to play and listen to the directions on how to play them. Don't forget to make observations about each game as you play them or just afterward.

Part C: Analysis and written reflection (individually)

a Summarize the events that took place on Games Day.

b Calculate the experimental probabilities using the data on your activity sheet. How do those values compare to the theoretical probabilities you calculated originally when you created the game? How do you explain any differences?

c Which game was the most fun to play? Explain.

d Was your game as successful as you had hoped? Did your classmates enjoy playing it? What aspects of made it a success? What would you change if you could? Provide specific examples with justifications/evidence.

e What makes a game fun to play?

f Can winning be calculated, or is it luck? Explain.

③ Integers

Exploring Earth has also meant exploring the world of positive and negative numbers, as you will learn in this unit. But where else are integers used? Sport and money feature prominently in many people's lives but few people realise the role integers play.

🌐 Identities and relationships

Teams and competition

Wherever there is competition, there are comparisons. Being able to analyse and compare athletes allows teams to decide which players to keep and which ones to replace. Individual athletes use the same measures when comparing themselves to opponents.

Individual races, like those in track and field, swimming or skiing, use positive and negative numbers to compare times. A standard - such as the current world record - is set and times are judged against that standard. Slower times are given as positive values while those ahead of the standard are reported with negative values. -1 second means the athlete is one second faster than the standard.

In hockey, a player earns a +1 every time she is on the ice when her team scores a goal. -1 is earned for being on the ice when the opposing team scores a goal. This plus/minus statistic is used as a measure of a player's contribution to the team's performance.

The best plus/minus rating was recorded by Bobby Orr during the 1970-71 season (+124), while the worst rating (-82) was earned by Bill Mikkelson in 1974-75.

Globalization and sustainability

Markets and commodities

One way that people try to increase their wealth is by investing in the stock market or in commodities. The value of both stocks and commodities fluctuates, and these fluctuations can be described using intergers.

At what point would you buy or sell stock? Would you wait for the price to look as though it's increasing or would you try to time it when the price is low? Analysing prices and representing changes with integers can help to identify trends and influence your buying and selling strategy.

Rather than purchasing a single stock or commodity, some people *diversify* by investing in groups of them. This can take several forms, but the idea is that losses in one area are offset by gains in another. Diversification is attractive to people who want to minimize their risk whilst benefitting from the advantages investments offer over typical bank accounts. Once again, the use of integers can provide insight into the direction investments are taking.

.91	81,800	334	1,483	44.05	0.
.71	30,681,300	56,084	1,356	9.49	3.
.00	363,900	7,337	7,108	10.28	1.
.58	1,439,700	4,944	7,243	11.72	1.
.16	5,802,400	7,977	625	85.33	2.
.98	2,195,800	22,337	5,555	26.89	13
00	14,400	503	8,750	10.77	1.
.76	1,575,400	36,161	48,510	24.81	1.
.87	115,600	3,380	18,487	19.58	3

3 Integers
Human explorations

Related concepts: Quantity and Representation

Global context:

In this unit you will embark on a journey of vast proportions, to see how humans have been able to explore and describe our planet. As part of the global context **orientation in space and time**, you will learn that an understanding of numbers has been instrumental in being able to explore the Earth and the furthest reaches of outer space.

Statement of Inquiry:

Being able to represent different forms of quantities has helped humans explore and describe our planet.

Objectives

- Defining, comparing and ordering integers
- Defining and evaluating the absolute value of a number
- Performing the operations of multiplication, division, addition and subtraction with integers
- Applying mathematical strategies to solve problems involving integers
- Plotting points on the Cartesian plane

Inquiry questions

F What is a quantity?

C How are different forms of quantities represented?

D How do we define where and when?

ATL1 Transfer skills

Make connections between subject groups and disciplines

ATL2 Reflection skills

Consider personal learning strategies

You should already know how to:

- round numbers
- evaluate numbers with indices/exponents
- calculate basic probability
- evaluate expressions using the order of operations

Introducing integers

When you first learned to count, you relied on the numbers 1, 2, 3 and so on. That was all you needed in order to count or keep track of things. Later on, you learned that there are numbers in between these values that represent parts or portions of a whole, and that can be written in fraction or decimal form. In this unit, you will expand that knowledge of number systems to include a set of numbers called *integers*.

The development of numbers and number systems seems to coincide with our need for them. Zero wasn't a recognized value until it became necessary as a placeholder for 'nothing'. Fractions and decimals were used once we realized that we didn't always need to deal in whole amounts of items. So, what other numbers could there be? What purpose do they serve? As it turns out, numbers have been instrumental in helping us to navigate and describe our planet and outer space. Being able to perform operations with these values may even lead you to the next great human exploration.

Integers

What is an integer?

Numbers, like words, can have opposites. You are already familiar with opposite words in English, like those in the activity below.

ATL1

Activity 1 – Opposites

1 Match each of the words below with its opposite.

strong   subtract   over
natural
same   weak
near
negative
always
distant   different
artificial   under   never   uncertain
positive   absolute   add

2 Create five other pairs of words that are opposites. As a class, share them and see who can create a unique pair.

3 Give the opposite of each of the following words:

   a hot   **b** rising   **c** up   **d** west   **e** minus

4 What do you think is the opposite of 7? Explain.

The opposite of a positive number is a negative number. For example, the opposite of 5 is written as −5 (pronounced 'negative five' or 'minus five'). The set of *integers* is the set of all positive and negative numbers (and zero) that are neither fraction nor decimal.

Integers can be represented on a number line, like the one below:

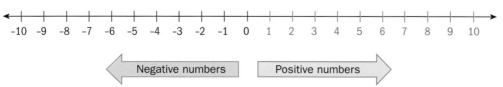

-10 -9 -8 -7 -6 -5 -4 -3 -2 -1 0 1 2 3 4 5 6 7 8 9 10

Negative numbers   Positive numbers

A number line can be drawn vertically as well:

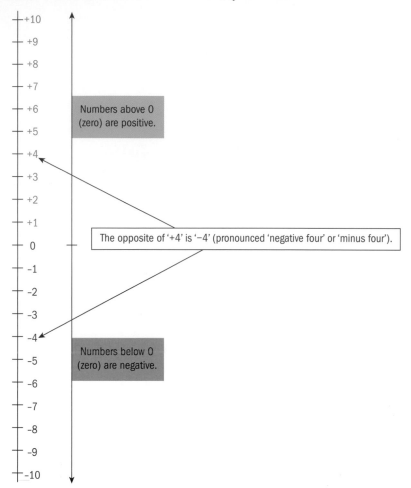

The opposite of '+4' is '−4' (pronounced 'negative four' or 'minus four').

Numbers above 0 (zero) are positive.

Numbers below 0 (zero) are negative.

Note that positive numbers can be written with or without the '+' sign. For example, '8' and '+8' are equivalent. Frequently the + sign is left out, and you can assume that the number is positive.

Reflect and discuss 1

- What quantity on a number line is always in the middle of two opposite values? Explain.

- On the horizontal number line, where are the larger quantities? Where are the smaller ones? How do you know?

- On the vertical number line, where are the larger quantities? Where are the smaller ones? How do you know?

- Which representation (the horizontal number line or the vertical number line) do you prefer? Explain.

Practice 1

1 State whether or not each number below is an integer.

a $\frac{2}{3}$ b −11   c 5.6   d 24

e −1   f $\frac{1}{2}$ g 0   h −6

i 8   j −7.8   k −34.25   l 999 999

2 Write down the opposite of each integer. Represent them both on a number line, using an appropriate scale.

a −12   b 7   c −82

d −3   e 26   f −103

3 Insert the appropriate symbol (< or >) between each pair of quantities to indicate which one is greater.

a −4   5   b 7   −10   c −4   −8   d 0   −3

e 9   6   f −12   −20   g −1   0   h −52   −67

i −100   −14   j −8   −2   k 3   −2   l −6   −31

ATL1

4 Stars and planets in the night sky have been used as a tool for navigation since ancient times, and scientists believe that some animals actually use the stars as a guide. In approximately 129 BC, the Greek astronomer Hipparchus created a star catalogue in which he described the 'stellar magnitude', or brightness, of a star as seen from Earth (now called *apparent magnitude*). Later, astronomers modified the system so that the magnitude of each object in the sky was represented by a single quantity. Magnitudes of objects in the night sky are given below.

Object	Stellar magnitude
Our sun	−27
The full moon	−13
Venus	−4
Jupiter	−2
Sirius (star)	−1
Halley's comet	+2
North star (Polaris)	+2
The Andromeda galaxy	+3
Uranus	+5

▶ Continued on next page

a Some people argue that this system makes little mathematical sense. Explain why, using specific examples.

b Name two pairs of objects that have opposite magnitudes. What do you think this means, if anything? Explain.

c Why do you think this system was created? Explain.

5 Insert the appropriate symbol ('+' or '−') in order to make the statement true.

   a □ 5 > −4   **b** □ 3 < −1   **c** −7 < □ 10

   d −2 > □ 5   **e** 3 < □ 4   **f** □ 12 > 9

6 Write down the following numbers in order from greatest to least.

   a −3, −1, −7, 9, −5   **b** 0, 8, −2, −12, 4, −4, −20

   c 2, −5, −9, −3, 0, 4, −10   **d** −45, −31, −52, −67, −34, −49

   e −104, −112, −101, −121, −134, −125, −119

ATL1   **7** When describing locations on Earth, their altitude above or below the ocean (*sea level*) is represented by an integer. Altitudes above sea level are represented by positive numbers while those below sea level are represented by negative values. Sea level is considered as 'zero'.

The table below gives the altitude of various locations on Earth.

Location	Altitude (m)
Mount Everest	+8850
Marianas Trench	−11 000
Baku, Azerbaijan	−28
Paris, France	+34
Manila, Philippines	+7
Cusco, Peru	+3399
Death Valley, USA	−85
Amsterdam, Netherlands	−2
Nairobi, Kenya	+1728
Deep Lake, Antarctica	−50

a Write down these locations in order from lowest to highest altitude.

b Place all of these locations on a number line that is roughly to scale.

c Research five more locations, including your hometown, and find their altitude above or below sea level. Place each of them in the appropriate spot in your list and on your number line.

▶ Continued on next page

If you want to find the altitude of any location on Earth, visit whatismyelevation.com and choose your spot!

d Mount Everest has the highest altitude on Earth, while the Marianas Trench (the deepest part of the Earth's oceans) has the lowest altitude, and both have been explored by humans. How does the depth of the Marianas Trench compare to the altitude of Mount Everest? Explain.

e Research and describe the similarities between conditions at extreme altitudes, above and below sea level.

f New Orleans is a city in the United States that is below sea level. Research five places on the continent where you live that are below sea level. What are the possible disadvantages of living in such locations?

Absolute value

On any voyage of discovery, it is important to take into account the temperature of the region to be explored. The equipment that was necessary to discover the New World was very different than that required to explore Antarctica or even the Moon. While there are several units of temperature, the Celsius scale is the most common. A thermometer has both positive and negative numbers, representing temperatures above and below zero.

In some areas, the temperature year-round is relatively stable. In others, the swing between winter and summer temperatures can be almost unbelievable. Verkhoyansk in eastern Russia, for example, has had temperatures as low as $-68\,°C$ and as high as $37\,°C$!

Reflect and discuss 2

- On a thermometer or number line, what value is in the middle of the positive and negative integers?

- Is zero positive or negative? Explain. (Feel free to conduct research before answering this question.)

- How does the temperature measured from zero compare for a number and its opposite? Explain using an example.

A number and its opposite are both the same distance from zero. This distance from zero is called the *absolute value* of a number and is represented using the symbol '| |' . For example, |−3| is 'the absolute value of three'.

| −3 | = 3, because −3 is three units from zero.

Investigation 1 – Absolute value

criterion B

1 Copy the table and fill in the missing values.

Absolute value	Distance from zero (units)	Conclusion
|−4|		|−4| = ?
|−10|		|−10| = ?
|3|		|3| = ?
|−7|		|−7| = ?
|12|		|12| = ?
|−31|		|−31| = ?
|75|		|75| = ?

2 What generalization can you make about the absolute value of any quantity? Explain.

3 Verify your generalization with another example of your choosing.

4 Justify why your generalization works.

Reflect and discuss 3

- How do the absolute values (magnitudes) of opposite integers compare? Explain.

- Nghia says, 'The absolute value of a number is simply the number without its sign (+ or −).' Does this show an understanding of absolute value? Explain.

ATL2
- How difficult is the concept of absolute value for you?

- What can you do to become an even more efficient and effective learner?

Did you know?

In 1742, Anders Celsius proposed the Celsius scale that we use today for temperature. His system was based on the freezing point and boiling point of water, with one hundred divisions between them. However, in order to avoid negative numbers, he initially proposed the freezing point to be 100 degrees and the boiling point to be 0 degrees. The quantities were eventually reversed.

Practice 2

1 Evaluate the absolute value of each of the following:

   a $|-2|$ **b** $|12|$ **c** $|8|$ **d** $|-7|$ **e** $|-34|$

   f $|102|$ **g** $|-79|$ **h** $|41|$ **i** $|-11|$ **j** $|0|$

2 Insert the appropriate symbol (<, > or =) between each pair of quantities.

   a $|-7|$ 3   **b** 4 −6   **c** $|4|$ $|-6|$ **d** $|-8|$ $|8|$

   e $|-10|$ $|-3|$ **f** $|-14|$ $|-19|$ **g** 0 −5   **h** $|0|$ $|-5|$

   i $|-100|$ $|-140|$ **j** $|12|$ $|22|$ **k** 7 $|11|$ **l** $|16|$ $|-16|$

3 Fifth Avenue in Manhattan, New York, divides the city into east and west. If you are on 41st Street, you are on either East 41st Street or West 41st Street, depending on which side of Fifth Avenue you are.

   a If you were to represent East 41st and West 41st streets on a number line, what would Fifth Avenue represent? Explain.

   b Assuming all city blocks are the same length, how should the distance between 231 East 41st Street and Fifth Avenue compare to the distance between 231 West 41st Street and Fifth Avenue? Explain.

   c Explain how you could use integers in addresses, instead of having East 41st and West 41st streets.

   d How does the address of a house on East 41st Street relate to the concept of absolute value? Explain.

   e Suggest two reasons why East and West streets are used instead of using negative numbers for addresses.

▶ Continued on next page

4 The lowest place in the southern and western hemisphere is the Laguna del Carbon in Argentina, which is at an altitude of 105 meters below sea level.

a Represent '105 meters below sea level' as an integer.

b How do altitudes reported 'above' or 'below' sea level relate to absolute value? Explain.

c Find as many words as you can that represent 'positive' and 'negative'. Fill in a table like this one, and share your results so that the class as a whole can create a thesaurus of words that represent 'negative' and 'positive'. Add more rows as necessary.

Words that represent 'positive'	Words that represent 'negative'
above	below

ATL1 **5** Roald Amundsen was an explorer from Norway who explored both the Arctic and the Antarctic, and was the first person to travel to the South Pole. In a race against explorer Robert Falcon Scott, Amundsen's expedition left Kristiansand, Norway, in August 1910 and arrived at the South Pole in December 1911, about one month before Scott.

Along his voyage, Amundsen experienced a wide range of temperatures at locations like those in the table below.

Location	Date	Average monthly temperature (°C)
Kristiansand, Norway	Aug 3, 1910	+15
Funchal, Portugal	Sept 9, 1911	+22
Whale Bay, Antarctica	Jan 14, 1911	−2
Framheim, Antarctica	Feb 14, 1911	−11
South Pole	Dec 14, 1911	−28

a Arrange the locations according to their temperature from lowest to highest.

b Arrange the temperatures from lowest absolute value to highest absolute value.

c What temperature has the same absolute value as that of Funchal? Research a location on Earth with that average temperature.

d Write down two positive temperatures and two negative temperatures that would have an absolute value between those of Framheim and Kristiansand.

The coordinate grid

In order to define a location, mathematicians use a *coordinate grid* called the *Cartesian plane*. Named after the mathematician René Descartes, it includes both a horizontal and a vertical number line as seen below. The horizontal line is called the *x-axis* and the vertical line is called the *y-axis*.

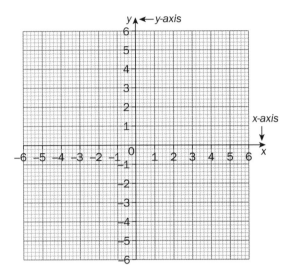

A location on the grid is called a *point*. Points are usually named with capital letters and their position is given using *coordinates*, as you will see in Investigation 2.

Did you know?

While the Cartesian plane is named after René Descartes, a similar sort of system was first used by a Greek geographer, philosopher and mathematician named Dicaearchus. He created a map using two reference lines, one going through the city of Rhodes and the other through the Pillars of Hercules. Eratosthenes, another Greek philosopher and geographer, used a similar map, though he added lines parallel to the original reference lines. This system is similar to our current system of latitude and longitude.

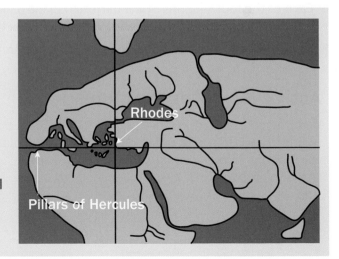

Investigation 2 – Plotting coordinate points

When you draw points on a coordinate grid, this is referred to as *plotting points*. Several points have been plotted below. Answer the questions to determine how to name the coordinates of a point.

Some of the points on the grid below are described using their coordinates:

$$A(-4, 3) \qquad Q(0, 4) \qquad W(4, -4)$$

1 Based on the given coordinates, write down the coordinates of the remaining points.

$$P(\ ,\ ) \quad N(\ ,\ ) \quad E(\ ,\ ) \quad S(\ ,\ ) \quad M(\ ,\ ) \quad J(\ ,\ ) \quad B(\ ,\ )$$

2 Describe how to determine the coordinates of a point on the Cartesian plane.

3 What do all of the points on the x-axis have in common? What do all of the points on the y-axis have in common?

4 Research the name given to the point where the two axes meet and write down its coordinates.

5 Which quadrant has negative values for both the x and y axes?

6 Each of the four sections of the Cartesian plane (called *quadrants*) is represented with a number. Do some research and write down how the quadrants are numbered.

7 On the coordinate grid, what do negative numbers represent?

Activity 2 – Integer pictures

Using a coordinate grid ranging from −10 to 10 on both axes, plot the points in each group of ordered pairs below and join them up in order, using a ruler to make straight lines. Each group is its own shape. What is your picture of?

1 (0, 0), (6, 0), (0, 8)

2 (−1, 0), (−5, 0), (−1, 6)

3 (−5, −2), (6, −2), (5, −4), (−4, −4)

4 (0, 8), (0, 10), (2, 9)

▶ Continued on next page

5 On the same size grid, write your own instructions to create a simple picture like the one above. Check your instructions for accuracy of plotted points, then hand in the instructions to your teacher. Your teacher will redistribute the instructions to the class and you will draw each other's pictures.

Practice 3

1 Write down the coordinates of each of the points in the coordinate grid below. Name the quadrant in which each is located.

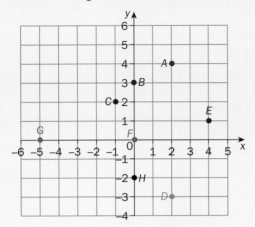

2 Create a Cartesian plane and plot the following points.

    a (−1, 7)   **b** (0, 5)   **c** (−2, −2)   **d** (1, 0)

    e (−3, 4)   **f** (3, −4)   **g** (3, 6)   **h** (−5, −3)

ATL1

3 When exploring what is located under the Earth, archeologists and paleontologists use a coordinate grid system. In order to be able to locate and identify artefacts or fossils once they have been removed, their location in terms of coordinates is recorded. In this photograph of a dig in Guadalajara, Spain, which unearthed mammoth fossils, how do you think paleontologists would use coordinates to identify the positions of the bones? Give a specific example, using the bone highlighted in blue.

Activity 3 – Battleship

Pairs

This activity is based on the popular board game *Battleship*.

You and your partner will each create two grids, ranging from −10 to 10 in both x and y directions, like the ones below. Label one grid 'My battlefield', and the other 'Opponent's battlefield:

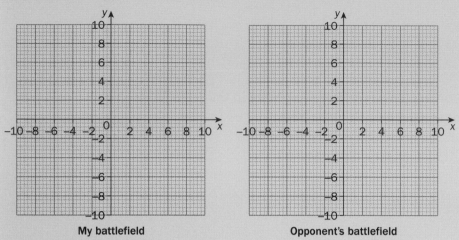

My battlefield   **Opponent's battlefield**

Keeping your grids hidden from your opponent, you will place your five battleships anywhere on the My battlefield grid. A battleship must be placed on consecutive coordinates and can be horizontal, vertical or diagonal.

The battleships are:

- An aircraft carrier (made up of 5 coordinate points)
- A battleship (made up of 4 coordinate points)
- A submarine (made up of 3 coordinate points)
- A destroyer (made up of 3 coordinate points)
- A patrol boat (made up of 2 coordinate points)

One person starts by calling out a coordinate point. The other person will reply with either 'hit' or 'miss'. If it is a hit, the person who called the point will place an X on that coordinate on their copy of the Opponent's Battlefield grid. If it is a miss, a dot is placed on that point.

Partners take turns calling points.

The first to sink all battleships wins the game!

Operations with integers

Integers can be compared and their absolute value (or magnitude) found. As with other forms of numbers, you can perform all of the mathematical operations with integers. How do these operations compare to operations with positive numbers or even decimals and fractions?

Multiplication and division

In order to explore the Moon, the Saturn V rocket carrying Apollo 11 needed to overcome the force of gravity, the force of attraction between the Earth and every object on it. Experimentation helped scientists figure out that the force of gravity (measured in newtons) can be approximated by multiplying the mass of an object, measured in kilograms, by −10 (negative since the force of gravity is downward). If the mass of the rocket (including payload and fuel) was almost 3 000 000 kg, approximately how large was the force of gravity on it?

What does it mean to multiply by a negative number? How is it similar to or different from multiplying and dividing by positive numbers? The next investigation will help you figure that out.

Investigation 3 – Multiplying positive and negative numbers

criterion **B**

1 Copy each equation and fill in the missing values.

$4 \times 4 = $ ___

$4 \times 3 = $ ___

$4 \times 2 = $ ___

$4 \times 1 = $ ___

$4 \times 0 = $ ___

2 What pattern do you notice in your answers? Explain.

3 Use the pattern to predict the following:

$4 \times (-1) = $ ___

$4 \times (-2) = $ ___

$4 \times (-3) = $ ___

$4 \times (-4) = $ ___

▶ Continued on next page

4 Write down a rule in words about multiplying a positive number and a negative number, by filling in the blanks: 'I think _____ because _____.' Compare your rule with that of a peer.

5 Verify your rule by exploring a similar pattern, this time starting with 10×5 and ending at $10 \times (-4)$. Does your rule still work? Explain. If you need to, make changes to your rule and verify it with another pattern of your choice.

6 Suppose you represented negative integers with red tokens or chips (●). Draw a representation of $4 \times (-3)$. Describe how this representation verifies your rule for multiplying a positive integer by a negative one.

7 Why is $4 \times (-3) = (-3) \times 4$? Explain using one of the properties of mathematics. You may draw a representation using red tokens or chips to help you.

8 Copy the following and fill in the missing values.

$(-3) \times 4 = $ ____

$(-3) \times 3 = $ ____

$(-3) \times 2 = $ ____

$(-3) \times 1 = $ ____

$(-3) \times 0 = $ ____

$(-3) \times (-1) = $ ____

$(-3) \times (-2) = $ ____

$(-3) \times (-3) = $ ____

9 What pattern do you notice in your answers? Explain.

10 Write down a rule about multiplying two negative numbers by filling in the blanks: 'I think _____ because _____.' Compare your rule with that of a peer.

11 Verify your rule by exploring a similar pattern, this time starting with $(-2) \times 3$ and ending with $(-2) \times (-4)$. Does your rule still work? Explain. If you need to, make changes to your rule and verify it with another pattern of your choice.

12 Summarize the rules you have generalized regarding the product of positive and negative integers.

Reflect and discuss 4

- How are the answers to 6×7, $6 \times (-7)$ and $(-6) \times 7$ related? Explain.

- The force of gravity (in N) is more precisely calculated by multiplying the mass of an object (in kg) by -9.8. How large is the force of gravity on a mass of 8 kg? Explain.

You have generalized a rule for multiplying two integers. Does dividing follow a similar pattern? You will explore this in the next investigation.

Investigation 4 – Dividing by negative numbers

Negative integers are often represented with red tokens (●), and positive integers with black tokens (●).

1   Draw a diagram of how you would represent –6 with tokens.

2   Divide the tokens into three groups of equal size. How many tokens are in each group? How would you represent this as an integer?

3   Write down a division statement that represents what you just did, including the answer.

4   Repeat the same steps for the following examples:

    a $8 \div 4$ **b** $-12 \div 2$

    c $10 \div 5$ **d** $-14 \div 7$

5 Generalize what you have found so far for the division of an integer by a positive number.

Since this model does not allow you to divide tokens into a negative number of groups, you can try something a little different.

6   Represent –12 with tokens. Draw a picture of your representation.

7   Divide the tokens into groups of –3. How many groups of –3 can you make?

8   Write down a division statement that represents what you just did, including the answer.

9   Repeat the same steps for the following examples:

    a $-16 \div (-4)$ **b** $-15 \div (-5)$

    c $-20 \div (-10)$ **d** $-18 \div (-3)$

10 Summarize what you have found for the division of negative numbers.

Reflect and discuss 5

- You found that $-6 \div 3 = -2$. What do you think $6 \div (-3)$ is equal to? Explain.

- Are the rules for multiplying and dividing integers the same? Explain.

(Q) In physics, 'velocity' is a speed with a direction. It can be either positive (in this example, heading east) or negative (heading west). An airplane is traveling at −80 km/h down a runway.

a The pilot needs to triple her velocity in order to take off. What is her desired take-off velocity?

b Once in flight, the plane's cruising velocity will be −800 km/h. How many times faster than the runway speed (−80 km/h) is this?

(A) **a** $-80 \times 3 = -240$

> To triple means to multiply by 3.

> A positive number multiplied by a negative number results in a negative number.

The desired take-off velocity is −240 km/h, which can also be represented as 240 km/h in a westerly direction.

b $-800 \div -80 = 10$

> In order to find how many times faster it is, you need to divide.

> A negative number divided by a negative number results in a positive number.

The cruising velocity (−800 km/h) is 10 times the runway velocity (−80 km/h).

ATL2 **Activity 4 – Memory aid**

In order to remember the rules you just generalized, it is sometimes helpful to come up with a clever memory aid, like the one below (based on the Transformers toys/movies). Autobots are the good transformers and are seen as positive. Decepticons are the bad transformers and are seen as negative. Coming to Earth is defined as the positive direction, and leaving the Earth as the negative direction.

If the Autobots (+) come to Earth (+), that is a good thing (+).

If the Autobots (+) leave Earth (−), that is a bad thing (−).

If the Decepticons (−) come to Earth (+), that is a bad thing (−).

If the Decepticons (−) leave Earth (−), that is a good thing (+).

1 Create your own memory aid for the rules for multiplying and dividing integers.

2 As a class, share your memory aids and then pick the one you like the best!

You have generalized a rule for multiplying and dividing *two* integers. Do you think there is a rule for *any* number of integers?

Investigation 5 – Patterns with negative numbers

criterion **B**

1 Evaluate the following multiplications involving more than 2 integers:

a (−7)(+5)(−3)

b (−4)(−2)(−9)

c (+7)(+6)(+2)

d (+6)(−15)(+2)

e (−10)(+5)(−3)(−2)

f (+7)(+3)(+11)(−2)

g (−3)(+2)(+12)(−2)

h (−3)(−2)(−12)(−2)

i (+7)(+3)(+11)(−2)(+1)

j (+2)(+4)(+10)(−2)(−3)

k (−9)(−2)(+11)(+5)(−1)

l (−4)(−11)(−8)(−3)(−2)

2 Look at all of your results above. Can you see a pattern that relates the signs of integers in each expression to the sign in the overall answer? Write a general statement about the result of multiplying several integers.

3 Given your statement above, write a problem-solving process (steps) that can be used to multiply several integers together.

4 Verify your rule for a different example and justify why your rule works.

5 Do you think the same pattern would occur when you divide integers? Would this pattern also apply for a combination of multiplying and dividing integers? Justify your response by testing your pattern.

Reflect and discuss 6

- In language, what is the result of using one or more negative words?
    - 'I disagree with you.' (single negative)
    - 'I do not disagree with you.' (double negative)
    - 'I cannot say that I do not disagree with you' (triple negative)
- How does this relate to the rules for multiplying and dividing integers? Explain.

WEB LINK

Integers: Multiply and Divide is an interactive activity on the LearnAlberta.Ca website. You can consolidate your understanding of the patterns and rules for multiplying and dividing integers.

Investigation 6 – Exponents

In this investigation, the goal is to generalize a rule for what happens when you raise an integer to an exponent. However, this time you will create the procedure yourself.

1 Decide on a way of testing what happens when you raise positive and negative numbers to any exponent greater than zero.

2 Write down your procedure and carry it out. Be sure to include a table of results from which you will make your generalization. Your generalization must be made after seeing at least ten examples.

3 Write down the rule you generalized.

4 Verify your rule for another case and then justify your rule.

Reflect and discuss 7

- How does your rule from Investigation 6 relate to the rule you generalized for multiplying several integers? Explain.

- Is $-3^2 = (-3)^2$? Explain.

Practice 4

1 Find the following products.

a -4×8 **b** $-3 \times (-7)$ **c** $10 \times (-6)$ **d** $-5 \times (-9)$ **e** -11×4

f 8×6 **g** $-7 \times (-7)$ **h** -6×9 **i** $-2 \times (-1)$ **j** 7×5

k -11×11 **l** $9 \times (-3)$ **m** $-6 \times (-6)$ **n** 4×12 **o** -12×4

p $-5 \times (-2)$ **q** $2 \times (-12)$ **r** 10×7 **s** $-3 \times (-2)$ **t** -9×1

2 Match the expressions that have the same value.

-2×6 $10 \times (-2)$ $-3 \times (-4)$ 2×10

$-5 \times (-4)$ -12×1 -3×4 $-2 \times (-10)$

$-1 \times (-12)$ -20×1 -4×5

$-1 \times (-20)$ $-6 \times (-2)$ -4×3

▶ Continued on next page

3 Insert the value in the brackets that makes the statement true.

a $-5(\ ) = -30$ **b** $7(\ ) = -28$ **c** $-4(\ ) = 12$

d $3(\ ) = 27$ **e** $-8(\ ) = 56$ **f** $11(\ ) = -66$

4 Write down four different pairs of integers that multiply to give each of the following products. Make sure each pair contains at least one negative quantity.

a -24 **b** 36 **c** 30 **d** -32

5 a Find 3 numbers that multiply to give each of the products in question 4.

 b Find 4 numbers that multiply to give each of the products in question 4.

6 The table below gives the altitude of various locations on Earth.

Location	Altitude (m)
Danakil Depression, Ethiopia	−125
Baku, Azerbaijan	−28
Deep Lake, Antarctica	−50
Kristianstad, Sweden	−2
Laguna Salada, Mexico	−10
Salton Sea, USA	−80
Lammefjord, Denmark	−7
Caspian Sea depression, Kazakhstan	−40

a Order the locations from lowest altitude to highest.

b Find the absolute value of the altitude of Baku, Azerbaijan. What does this quantity represent?

c To solve each of the following, write down a mathematical expression and evaluate it.

 i How many times deeper is Deep Lake than Laguna Salada?

 ii If their shapes and surface dimensions were proportionally the same, how many times could the Salton Sea completely fill up the Caspian Sea?

 iii How many times lower is Baku, Azerbaijan than Lammefjord, Denmark?

 iv How many times lower is the Danakil Depression than Laguna Salada? Is this number an integer? Explain.

7 List all of the integers between −10 and 10 that are factors of each of the following.

a -24 **b** 36 **c** -60 **d** -100 **e** 64

▶ Continued on next page

8 Simplify these values.

a $(-6)(-11)$ **b** $(+12)(+7)$ **c** $(+18)(-3)$ **d** $(+84) \div (+4)$

e $(+56) \div (-7)$ **f** $(-49) \div (-7)$ **g** $(+13)(-4)$ **h** $(-16)(+7)$

i $(-15)(+6)$ **j** $(+5)(-20)$ **k** $(+8)(-14)$ **l** $(-11)(-7)$

m $(+64) \div (+2)$ **n** $(-84) \div (+7)$ **o** $(+60) \div (-6)$ **p** $(-2)(-3)(-4)$

q $(+6)(-2)(-3)$ **r** $(+5)(+12)(+7)$ **s** $(+2)(+9)(-3)$ **t** $(-3)^3$

u $(-2)^5$ **v** $(-5)^2$ **w** $(-1)^{43}$ **x** $(-10)^4$

9 a Determine the magnitude of the force of Earth's gravity on each of the following spacecraft used to explore outer space. Remember, the force of gravity on Earth (in N) can be approximated by multiplying the mass of the object (in kg) by −10.

Spacecraft/mission	Mass (kg)	Objective
Viking	636 497	Land on and explore Mars
Cassini–Huygens	943 050	Orbit and photograph Saturn and its moon Titan
Voyager 1	632 970	Explore the outer planets in the solar system and the outer heliosphere
Hubble telescope (launched from the shuttle Discovery)	259 233	Explore the universe; take pictures of planets, stars and galaxies
Apollo	2 970 000	Explore the Earth's moon

10 In order to explore the Marianas Trench, the deepest point on Earth, Jacques Piccard and Don Walsh used a submersible craft called *Trieste*. They had to reach a depth of −10 900 m in a craft that could reach speeds of −60 m/min. At this speed, how long would it have taken the *Trieste* to reach this depth? Show your working.

Did you know?

In 1589, Galileo proposed that two objects of different mass will hit the ground at the same time if, they are dropped from the same height. Whether or not he performed this experiment from the Leaning Tower of Pisa is still debated. However, during the Apollo 15 mission to the Moon, astronaut Dave Scott proved Galileo's theory correct by dropping a falcon feather and a hammer from the same height. Without any air resistance, they did land at the same time. Used because the lunar module's name was the *Falcon*, the falcon feather is still on the surface of the Moon today, along with the hammer!

criterion D

ATL1

Formative assessment 1

On July 16, 1969, the Apollo 11 mission launched from Cape Kennedy, Florida, and, four days later, landed on the surface of the Moon. Neil Armstrong and Edwin 'Buzz' Aldrin spent just under 22 hours on the Moon's surface before rejoining Michael Collins in the command module and landing back on Earth on July 24, 1969. The mission was more than a race to the Moon; it also included lunar exploration and setting up monitoring equipment. Since then, many missions have gone to the Moon (or near it) to collect a wide range of data.

1 The force of gravity on the Moon is significantly less than it is on Earth. In order to approximate this force, multiply the mass of the object by -2. Remember, on Earth, the force of gravity is calculated by multiplying the mass by -10.

a Find the force of gravity on each of the following items, both on Earth and on the Moon. Show your working in your own copy of this table.

Equipment	Mass (kg)	Force of gravity on Earth (N)	Force of gravity on the Moon (N)
Lunar module (*Eagle*)	15 103		
Neil Armstrong	75		
Command module	5557		
Samples of lunar rock	22		

b How many times larger is the force of gravity on Earth than the force of gravity on the Moon? Show your working.

2 On the surface of the Moon, temperatures and elevations vary widely. Since there is no ocean, elevations are given in terms of how far above and below the average radius of the Moon (1737.4 km) they are.

Location	Temperature (°C)
Equator (min)	−180
Equator (max)	+125
South pole	−238
Hermite crater (north pole)	−250
Far-side highlands	−50

Location	Elevation (meters above/below average radius)
Mons Ampere	+3300
Janssen crater	−2800
Highest point on Moon	+10 786
Fra Mauro	0
Aristoteles	−3300
Mont Blanc	+3800
Mare Crisium	−1800
Far-side highlands	+1800
Near-side highlands	−1400

> Forces are quantities that can be measured in a variety of units. The SI unit of force is the newton (N), named after Sir Isaac Newton, because of his study of motion.

▶ Continued on next page

a Arrange the locations on Earth from coldest to warmest.

b Arrange the locations on the Moon from highest elevation to lowest.

c Arrange the absolute value of the elevations from highest to lowest.

d Explain why two elevations have the same absolute value.

e How many times lower is the Janssen crater than the near-side highlands? Show your working.

f Using the data in the tables, create a sensible question that would require multiplying or dividing integers but where the answer is not an integer. Be sure to write the question and a complete solution.

Reflection

g How difficult has it been for you to learn to multiply and divide positive and negative numbers? What factors are important for helping you to learn well?

For HD videos from the Apollo 11 mission, go to **nasa.gov** and search for 'Apollo 11 videos'.

Addition and subtraction

In order to explore our solar system, spacecraft need to be able to accelerate and propel themselves in space. Because chemical fuels are too heavy, ion thrusters or ion engines are used that produce a high level of propulsion by accelerating ions towards the back of the spacecraft, thereby pushing it forward at speeds up to 320 000 km/h! But what is an ion?

Every neutral atom on Earth has the same number of positive charges (protons) and negative charges (electrons). This means that if the sum of the positive charges and the negative charges is zero, the atom is neutral (has no charge). An ion, however, has a total positive charge or a total negative charge, depending on whether there are more protons or electrons respectively. The most common ion used in ion thrusters is xenon, with 54 protons and 53 electrons. What is the total charge of a xenon ion? In other words, what is (+54) + (−53)? How do you add integers to find the charge of **any** ion?

Did you know?

Xenon ion discharge from the NSTAR ion thruster of Deep Space 1. Credit: NASA
Modern ion thrusters use inert gases for propellant, so there is no risk of the explosions associated with chemical propulsion. The majority of thrusters use xenon, which is chemically inert, colorless, odorless and tasteless. Other inert gases, such as krypton and argon, also can be used. Only relatively small amounts of ions are ejected, but they are traveling at very high speeds. For the Deep Space 1 probe, ions were shot out at 146 000 kilometers per hour (more than 88 000 mph).

Pairs

Activity 5 – Hot air balloon

1 Draw or copy a hot air balloon onto a piece of paper or print one out. Cut the image out and draw a horizontal arrow on the basket.

2 With a partner, copy and fill in a table like the following. Feel free to do some research if necessary.

Action	In what direction does the hot air balloon move?
add hot air	
add weight (e.g. sandbag)	
remove (jettison) weight	
remove hot air	

3 You are going to represent puffs of air and sandbags with integers. Which one will be represented by positive integers and which one by negative integers? Explain your reasoning.

4 Compare your results with another pair in the class. Your teacher will compile the results for the whole class.

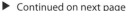
▶ Continued on next page

5 Draw a vertical number line that ranges from −10 to 10. Be sure to write down every single number in this range.

Suppose one sandbag and one puff of air each have the same effect on the hot air balloon. If you start at zero and add two puffs of air, you will go up two values. This can be represented as $0 + (+2) = 2$.

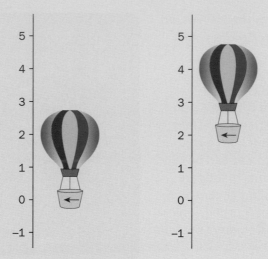

6 Now that your hot air balloon is at 2, what could you do to return to zero? Find two different ways and write your results using notation similar to the example. Represent these operations on your number line with your hot air balloon.

Adding and subtracting integers is a very important skill in mathematics. In the next investigation, you will use the hot air balloon and vertical number line from Activity 5 to generalize rules for these operations.

Investigation 7 – What goes up must come down

Pairs

When you are using your hot air balloon, note that the first number is the starting point. The symbol that follows it indicates whether you are adding or taking away. The second number represents *what* you are adding or taking away, either a sandbag or a puff of air.

▶ Continued on next page

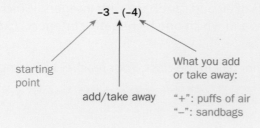

criterion B

1 In pairs, find the value of each of the following using your hot air balloon. Be sure you agree on the answer before moving on to the next one.

    a $(-3) - (-4)$ **b** $(+2) + (-5)$ **c** $-1 + (+7)$ **d** $3 - (+8)$

    e $-2 - (-1)$ **f** $-5 + (+3)$ **g** $6 - (-2)$ **h** $(-10) + (+5)$

    i $(+7) + (-6)$ **j** $9 - (+11)$ **k** $(-8) - (+2)$ **l** $(+4) + (-9)$

    m $(-9) - (-3)$ **n** $(+8) - (+15)$ **o** $(+6) + (-4)$ **p** $(-7) + (+6)$

    q $5 + (-4)$ **r** $-2 + (-6)$

2 In which questions did the hot air balloon go up? Explain.

3 If you could represent 'up' with either a positive sign or a negative sign, which would you choose? Explain.

4 In which questions did the hot air balloon go down? Explain.

5 If you could represent 'down' with either a positive sign or a negative sign, which would you choose? Explain.

6 Based on whether the balloon goes up or down, rewrite each of the questions in step **1** using only one sign between the two values. For example: $(-3) - (-4)$ can be rewritten as $-3 + 4$, since the balloon went up.

7 Verify your results in step **1** by simply going up or down from the starting value. Were your initial results always correct?

8 Rewrite each of the following with a single sign, and then find its value by simply going up or down from the starting value.

    a $6 - (+3)$ **b** $(-5) - (-11)$ **c** $(-4) - (+4)$ **d** $(+8) + (-15)$

    e $-3 - (-7)$ **f** $(+4) + (+5)$ **g** $-7 + (-3)$ **h** $1 - (+8)$

    i $(-6) + (+1)$ **j** $(+3) - (-6)$ **k** $-2 + (+4)$ **l** $9 + (-18)$

9 Write general rules for how to add and subtract integers. Use correct mathematical vocabulary and notation.

WEB LINK

Consolidate your understanding of the concepts in this investigation by heading to the **National Library of Virtual Manipulatives website** and searching for the 'color chips' activities for addition and subtraction of integers. Here, you will use color chips to illustrate addition and subtraction of integers.

Reflect and discuss 8

Reflect and discuss 8

- Fill in a table like the following to compare the results of operations with integers.

1st quantity	2nd quantity	Sign of the product of two quantities (+/–/unsure)	Sign of the quotient of two quantities (+/–/unsure)	Sign of the sum of two quantities (+/–/unsure)	Sign of the difference of two quantities (+/–/unsure)
+	+				
–	–				
+	–				
–	+				

- Do the rules for the product and quotient of integers also apply to their sum and difference? Explain.

- What aspects of addition and subtraction of integers use rules similar to those of multiplication and division? Why do you think this is?

Activity 6 – Two dice

Divide up the class into groups of four students. Each group will get two dice of different colors. Decide which color represents positive numbers and which one represents negative numbers.

Each person takes a turn rolling both dice and adding the integers together. The roller can decide to stick with that answer, or opt to roll again. Each person has just two chances to roll the dice. The winner is the person with the largest answer (which is +5 if you are rolling two standard dice).

Analysis

1 Write down each of the possible outcomes (totals) of rolling both dice.

2 Write down the totals in order from most likely (highest probability) to least likely (lowest probability).

3 On the basis of your findings, determine a strategy that you think will help you be the most successful at the game. Justify your strategy using mathematics.

4 Play the game using your strategy and comment on its effectiveness.

ATL1 **Example 2**

Q **a** What is the charge of an ion that has 10 electrons and 8 protons?

b How many electrons does a +3 ion have if it has 32 protons?

A **a** $(-10) + (+8)$

> Add the number of positive and negative charges.

$= -10 + 8$

> The two positive signs can be replaced with a single positive sign.

$= -2$

> Add. Use a model if necessary.

The charge is -2.

b $+3 - (+32)$

> If you subtract the number of positive charges from the total charge, you will find the number of negative charges.

$= 3 - 32$

> The positive and negative sign can be replaced with a single negative sign.

$= -29$

> Subtract. Use a model if necessary.

There are 29 electrons.

Another way to represent the addition and subtraction of integers is with red and black tokens. Using the tokens relies on the concept of the *additive inverse*. The additive inverse of a number is its opposite, because the *sum of a number and its additive inverse is always zero*.

$(-1) + (+1) = 0$

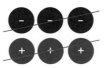

$(-3) + (+3) = 0$

Investigation 8 – Adding and subtracting

criterion **B**

1 Represent $(-4) + (+2)$ with two rows of tokens. Draw the representation in your notebook.

2 Circle or cross out the pairs that sum to zero, if there are any. What quantity is left?

3 Write down the operation and its result using integers.

▶ Continued on next page

4 Repeat this procedure for each of the following sums.

a $3 + (-7)$ **b** $(-3) + (+5)$ **c** $6 + (-5)$

d $-8 + (-1)$ **e** $-5 + 9$ **f** $1 + (-9)$

g $(+2) + (-3)$ **h** $(+4) + (+3)$ **i** $(-1) + (1)$

j $-7 + (+6)$ **k** $8 + (-5)$ **l** $-2 + (-2)$

5 Write down a generalization for the process of adding two integers.

Subtracting integers can be difficult with tokens. However, subtracting an integer is the same as adding its opposite. For example, for the hot air balloon, taking away air (subtracting a positive) resulted in the same motion as adding a sandbag (adding a negative).

$$3 - (+4) = 3 + (-4) \qquad -2 - (-5) = -2 + (+5)$$

So, when subtracting integers, simply rewrite it as an addition and then use the tokens.

6 Represent the following operations by adding the opposite integer.

a $(-3) - (-6)$ **b** $4 - (+5)$ **c** $-1 - (+7)$

d $(+8) - (+2)$ **e** $(-5) - (-2)$ **f** $-7 - (+3)$

g $+1 - (-5)$ **h** $-2 - (+1)$ **i** $-9 - (-11)$

j $3 - (+2)$ **k** $(+4) - (-1)$ **l** $-6 - (-7)$

7 Write down a generalization for the process of subtracting two integers.

ATL2 **Reflect and discuss 9**

- How do the absolute value of a number and the absolute value of its additive inverse compare? Explain.

- How do the rules you generalized for adding and subtracting integers with tokens relate to the rules you generalized with the hot air balloon?

- Which method do you prefer: the hot air balloon or the tokens? Explain.

WEB LINK

'Integers – add and subtract' is an interactive activity on the LearnAlberta.Ca website. You can consolidate your understanding of the patterns and rules for adding and subtracting integers.

Activity 7 – Go integer fish

This is a game for 3–5 players.

Take a standard deck of playing cards and remove the face cards (jacks, queens and kings). Each remaining card has a numeric value, with aces equal to 1. Red cards represent negative numbers and black cards represent positive numbers.

Each person is dealt five cards. The remaining cards are placed in the middle of the table and are called 'the lake'.

The goal is to collect pairs of cards that have a given sum that is set at the beginning of each game. (The first sum will be zero.)

The first player asks any other player for a specific card by saying, 'Do you have a _____ _____ ?'
(color) (value)

For example, 'Do you have a red two?'. If the player has that card, he/she gives it to the player who asked. The player must put on the table the pair of cards that total zero and then he/she may ask another player for a card in the same way. The first player continues to ask for cards and puts down 'zero pairs' until someone does not have the requested card. When that happens, the response is 'Go fish' and the player who asked retrieves a card from the lake. That player's turn is over.

Play continues with the next player asking a similar question to any other player, until he/she is told to 'Go fish'. If, at any time, a player has no cards left, then he/she takes a new set of five cards from the lake (or as many as remain in the lake, if there are less than five). Play ends when one player has used up all of their cards and there are no cards left in the lake. The winner is the person with the most cards that add up to the given total.

Play the first round with a sum of zero, then try a total of '−1' or '+2'. Remove the cards that are the same as the required total (e.g. if the total is −2, remove the red 2's). For more of a challenge, allow players to use any of the four mathematical operations.

Practice 5

1 Write down the operation indicated by each representation and then evaluate it.

a

b

▶ Continued on next page

c

d

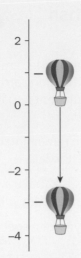

2 Evaluate these additions and subtractions.

a −2 + (+3) b −1 − 5 c 2 − (+7)

d 4 − (−2) e −8 − (+5) f −10 − (−12)

g 3 − 11 h −3 + 9 i 15 − 24

j −15 − 24 k −5 + (−8) l 2 + (+7)

m (−6) + (+10) n 5 + (−9) o −1 − (+1)

p 17 − (+21) q −15 − (− 4) r −3 + (−8)

3 Sum each integer pair, and then write down the sums in order from the lowest to the highest.

−11, 7   5, −3   −2, −4   −15, 18   7, −14   −8, 12   2, −1

4 Fill in the missing integer in each case.

a −3 + ( ) = −4 b −5 − ( ) = 7 c 2 + ( ) = −4

d 1 + ( ) = −9 e −10 − ( ) = −12 f 4 − ( ) = −4

g 9 + ( ) = 12 h −7 − ( ) = −2 i −2 + ( ) = −4

j 6 − ( ) = −3

5 a During the Age of Exploration, Jacques Cartier was the first European to reach the city that is now called Montreal in Canada. He arrived in October 1535 when the average temperature is 9 °C. In February, the average temperature is −8 °C. What is the difference in these temperatures?

b In Montreal, the temperature was 4 °C in the morning. Since then, the temperature has dropped 22 °C. Find the current temperature.

▶ Continued on next page

6 The average early morning temperature in Churchill, Manitoba, is −4°C in October. In January, it is 27°C lower. What is Churchill's average early morning temperature in January?

7 The elevation of the highest point in New Orleans, Louisiana, is 5 m above sea level. The lowest point in New Orleans is 8 m lower than the highest point. What is the elevation of the lowest point in relation to sea level?

8 Calculate the difference in altitude between the top of Mt Everest (8848 m) and the bottom of the Marianas Trench (−10 994 m).

9 **a** The magnitude of the Sun is −27. The magnitude of the Moon is −13. What is the difference in their magnitudes? Be sure to take the absolute value of your answer, since a difference cannot be negative.

   b Edwin claims that when two stars collide their magnitudes are added to produce the magnitude of the new star. Give an example that might support Edwin's claim and one that would help disprove it.

10 A submarine is situated 800 meters below sea level. If it ascends 250 meters, find its new position.

11 Eva Dickson was an explorer and the first woman to cross the Sahara Desert by car. Temperatures there can reach as high as 49°C. The coldest temperature recorded in her native Sweden is −53°C. Find the difference between these two temperatures.

12 In ion thrusters, it is rare to use ions with a charge other than +1, since a larger charge produces less propulsion. Rank the following ions in terms of their ability to produce the most propulsion. Justify your ranking with a mathematical operation involving integers.

   a 34 electrons, 37 protons   **b** 12 protons, 8 electrons

   c 8 electrons, 9 protons   **d** 25 protons, 20 electrons

   e 19 protons, 17 electrons   **f** 32 electrons, 26 protons.

Formative assessment 2

ATL1

In order to describe where we are on the planet, a system of latitude and longitude was established. Lines of latitude go around the Earth, parallel to the equator. Lines of longitude go around the Earth, but travel through the north and south poles. Measurements in this system are made in degrees, with northern latitudes being positive and southern ones being negative. Lines of longitude are divided by the Prime Meridian, which runs through Greenwich, England. East of this line is considered positive, while west of this line is negative.

James Cook was a famous explorer who went on several voyages of discovery. He measured and recorded both latitude and longitude using a sextant and some advanced mathematics (something that is much more easily done these days with GPS). On his first voyage, Cook traveled to the following locations:

Location	Date	Latitude (degrees)	Longitude (degrees)
Plymouth, England	August 1768	50	−4
Madeira, Portugal	September 1768	33	−17
Rio de Janeiro, Brazil	November 1768	−23	−43
Tierra del Fuego, Argentina	January 1769	−54	−70
Tahiti	April 1769	−18	−149
New Zealand	October 1769	−40	174
New Holland/Australia	April 1770	−25	134
Great Barrier Reef	June 1770	−18	148
Batavia, Java	October 1770	−6	106
Cape Town, South Africa	March 1771	−34	18
Kent, England	July 1771	51	1

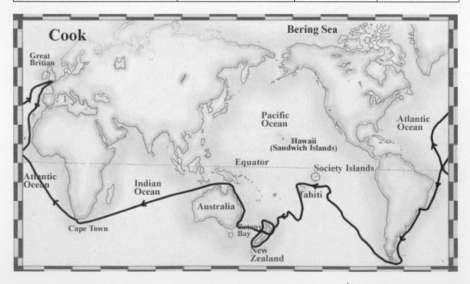

▶ Continued on next page

1 Find the distance he traveled between each location (in degrees) for both latitude and longitude. Show your working.

2 What does a negative answer in step **1** indicate? Explain.

3 What is the difference (in degrees) between his starting point and his ending point?

4 If you add all of the differences for latitude and then for longitude, do you get the difference found in step **3**? Justify your answer.

Reflection

ATL2

- How difficult has it been for you to learn to add and subtract integers?

- Is it easier or harder to add and subtract than multiply and divide integers? Why do you think that is?

- What can you do to become a more efficient and effective learner?

Multiple operations

Jacques-Yves Cousteau was probably the most famous ocean explorer ever. He helped create the original scuba gear and, eventually, his own submersible craft capable of diving as far down as 350 meters below sea level (−350 m). Once in the water, his craft would propel itself much like a squid, able to descend at a rate of approximately 50 meters every minute (−50 m/min).

Figuring out how long it takes to descend to −350 meters, or simply calculating the depth that should be reached after a specific time frame involves several mathematical operations. As with other number systems, integers follow the order of operations:

Brackets

Indices/Exponents

Multiplication/Division

Addition/Subtraction

Example 3

 Q If Cousteau's submersible craft sank to a depth of 5 meters below sea level and then the engine was used for 3 minutes, how far down would he be?

 A $-5 + 3(-50)$

> To calculate his depth, find how far he can dive in 3 minutes by multiplying the depth per minute by 3 minutes. Then, add that to his initial depth.

$= -5 - 150$

> The order of operations dictates that you must multiply first. After that comes addition/subtraction.

$= -155$

Cousteau would be 155 meters below sea level.

Activity 8 – Card game with integers

 Pairs

You will be put in pairs and each pair will be given a standard deck of playing cards. You will not use the picture cards, as you need only the numbers from 1 (ace) to 10. You can be put in groups of three with a referee if your teacher prefers.

- You are playing against one opponent at a time.
- Each person is dealt four cards.
- A red card represents a negative integer.
- A black card represents a positive integer.
- You can use any operation that you like – as many times as you like.
- You may use brackets, but no powers.

In each round you will be given a number, and the person who can get closest to that number using all four cards wins the round.

- Tie – one point
- Win – two points
- Achieves the exact number of the round – bonus one point.

 ## Practice 6

1 Evaluate each of these, where each involves multiple operations.

a $(-13) + (-4) \div (-2)$

b $(-15) + (+24)(-1)$

c $(-6 \times 3) - (-16) \div 4$

d $(-10 - 4)(-5)$

e $(+12) - (-4)^2$

f $14 \div (-7) + (-3)$

▶ Continued on next page

g $(+12) + (-9) + (+5) + (-2)$

h $(-5) - (+18) + (-9) - (-12)$

i $(-81) \div (+9)^2 + (-125) \div (-5)^2$

j $(+16) \div (-8) - (+48) \div (-4)$

k $(+6)(-8) \div (-2)^3$

l $(-50) \div (+5)(-9) \div (+10)$

m $+10 - (-54) \div (+6) + (-8)(-2) - (+5)$

n $+12 - (+4)(-7) + (-12)(-2) - (-18)$

o $\dfrac{(+6)(-2)(-4)}{(-9+6)(7-5)}$

p $\dfrac{(-18) + (-75) - (-24)}{(-4)(-8) - (+2)(-15) + 1}$

q $45 + (-72) \div (-6) + (-60) \div (-10)$

r $\dfrac{140 - (-160)}{68 - (-8)(+4)}$

s $\dfrac{(-20)}{-4} + (-68) - \dfrac{(-81)}{+9}$

t $\dfrac{15 - (+12) + (-14) + (+5)(-9)}{(-8)(2) - (-14)}$

2 While in space, it is not uncommon for astronauts to lose weight and return to Earth lighter than when they left. Suppose an astronaut's mass at the beginning of her mission was 96 kg. The first week she lost 2 kg, the second week she lost 4 kg, the third week she gained 2 kg, the fourth week she lost 1 kg and the fifth week she lost 3 kg. What was her mass after the five-week mission?

3 On 29 May 1953, Sir Edmund Hillary, a New Zealand explorer, and Tenzing Norgay, a Sherpa mountaineer from Nepal, were the first people to climb to the top of Mount Everest. Several attempts had been made since the first attempt in 1921, but climbing the mountain can be very dangerous. As you go up the mountain, which has an altitude of 8848 meters, the temperature drops and oxygen levels decrease, leaving explorers exposed to altitude sickness and hypothermia.

If the temperature when Hillary and Norgay left Kathmandu (+5000 m) was −8 °C, and the temperature drops 7 °C for every gain of 1000 meters, what temperature would they expect it to be at 8000 meters (Camp 4)? Using integers, write all of the calculations in a single step and evaluate.

4 Copy each of the following and insert operations and/or brackets to make the equation true. Do not use exponents/indices.

a −10   −6   2   −4   = 2

b 9   2   6   −3   = 13

c 7   4   −12   −2   3   = 27

Unit summary

The set of integers is the set of all positive and negative numbers (and zero) that are neither fraction nor decimal. Integers can be represented on a number line:

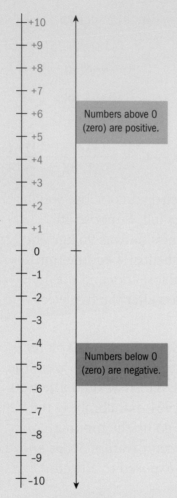

A number and its opposite are both the same distance from zero. This distance from zero is called the *absolute value* of a number and is represented using the symbol '| |'. For example, |−3| is 'the absolute value of three'.

$$|-3| = 3,$$ because −3 is three units from zero.

A coordinate grid (called the Cartesian plane) can be used to locate points in space. It is divided into four numbered quadrants.

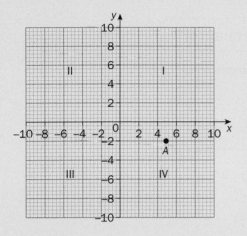

The horizontal line is called the *x*-axis and the vertical line is called the *y*-axis. A location on the grid is called a point. Points are usually named with capital letters and their position is given using coordinates.

For example, point *A* has coordinates $(5, -2)$.

The product or quotient of an even number of negative numbers is positive.

The product or quotient of an odd number of negative numbers is negative.

When adding or subtracting integers, try to simplify the question:

- 'subtracting a negative' or 'adding a positive' both result in going up, so replace this with a '+'.

  e.g. $-6 - (-2) = -6 + 2$

- 'subtracting a positive' or 'adding a negative' both result in going down, so replace this with a '−'.

  e.g. $-6 + (-2) = -6 - 2$

Once simplified, start at the first value and go up or down from there.

The additive inverse of a number is its opposite, because the sum of a number and its additive inverse is always zero.

You can also add integers by using tokens. Circle or cross out pairs of positive and negative tokens and report what is left.

When subtracting, add the opposite. For example, $-6 - (+2) = -6 + (-2)$. You can now use your tokens to add.

The order of operations must be followed when evaluating expressions : brackets, indices/exponents, multiplication/division, addition/subtraction.

Unit review

⬚ **Launch additional digital resources for this chapter**

Key to Unit review question levels:

Level 1–2   Level 3–4   Level 5–6   Level 7–8

1 Insert the appropriate symbol (< or >) between each pair of quantities to indicate which one is greater.

   a −8 ☐ −3   **b** −53 ☐ −62   **c** −7 ☐ −11

   d −1 ☐ 0   **e** |−5 | ☐ | −2| **f** |−11| ☐ |−12|

2 Insert the appropriate symbol ('+' or '−') in order to make the statement true.

   a ☐ 6 < −2   **b** ☐ 1 < 1   **c** 4 > ☐ 6   **d** 8 < ☐ 10   **e** −2 < ☐ 5   **f** ☐ 10 > −9

3 **Write down** the following quantities in order from least to greatest.

   a 0, −3, 2, −1, −7   **b** 9, −12, 5, −6, 7, −2, 1   **c** |−2|, |8|, |−4|, |−10|, |0|, |−1|

4 **State** the ordered pair (x, y) for each marked point on the grid below:

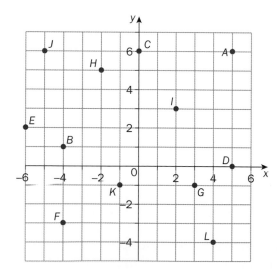

5 On graph paper, **plot** and **label** these points on a grid and **label** the x- and y-axes.

   E(5, 1)   F(0, 4)   G(−2, 6)   H(−4, −3)   I(4, −5)

   J(2, 7)   K(−5, 1)   L(0, −3)   M(6, 0)   N(1, −2)

   O(0,0)   P(−3, 5)   Q(−2, −6)

6   As seen previously, the magnitude of celestial objects uses positive and negative values, like those in the table below.

Object	Stellar magnitude
Our sun	−27
The full moon	−13
Venus	−4
Jupiter	−2
Sirius (star)	−1
Halley's comet	+2
North Star (Polaris)	+2
The Andromeda galaxy	+3

a   When their magnitude is as bright or brighter than that of Venus (−4), meteors are called 'fireballs'. Are fireballs 'meteors with a magnitude greater than −4' or 'meteors with a magnitude less than −4'? **Explain**.

b   **Explain** how the concept of absolute value makes this system more sensible.

7   The time of day at any one instant is different depending on where you are in the world. Earth is split up into different time zones, with Coordinated Universal Time (UTC) being considered zero. Locations to the east of UTC are *ahead* of UTC (positive) and locations west of UTC are *behind* UTC (negative). Some major cities and their time zones relative to UTC are given below:

Location	Time zone relative to UTC
Madrid, Spain	+2
Auckland, New Zealand	+12
Montreal, Canada	−4
San Diego, USA	−7
Tokyo, Japan	+9
Reykjavik, Iceland	0
Islamabad, Pakistan	+5
Istanbul, Turkey	+3
Rio de Janeiro, Brazil	−3

a Write down the cities as they would appear on a number line, based on their time zone relative to UTC.

b Which two cities have relative time zones with the same absolute value? What does this mean?

c Write down the cities in order of the absolute value of their relative time zone, from least to greatest.

d Do some research and find cities for which the relative time zone has an absolute value between that of Montreal and San Diego.

⑧ Evaluate.

a $(-3)(-5)$

b $-4 - 7$

c $(-5) + (+2)$

d $8(-11)$

e $2 - (+9)$

f $-3 - (-1)$

g $-24 \div 6$

h $-21 + (-12)$

i $(-4)(-2)(3)$

j $-18 \div -2$

k $8 - (-10)$

l $(+6)(-7)$

m $-2 - 8$

n $-10 \div -16$

o $(-3)(2)$

p $35 \div -7$

q $-14 + 10$

r $17 - 23$

s $-3(7)$

t $(+8) - (+13)$

u $(-3) + (-8) - (-5)$

v $(-2)(3)(-5)$

w $-12 - (-4) + (-17) - (+8)$

x $(-24) \div (-2) \div (-3)$

y $-7 + (-5) - (-6) - (+10)$

⑨ **Explain** the difference between $(3, -5)$ and $(-5, 3)$ and **describe** the way you would plot both points on a grid.

10 On a trip from 1903 to 1906, Roald Amundsen became the first person to sail through the Northwest Passage, the route across the Arctic that joins the Atlantic and Pacific oceans. His route is summarized in the table below and shown in the map that follows.

Location	Date	Average low temperature that month (°C)
Christiana, Norway	June 1903	+11
Godhaven, Greenland	July and August 1903	+2
Beechey Island, Nunavut	August 1903	0
King William Island, Nunavut	Winter 1903 – summer 1905	−37 (winter) +3 (summer)
Victoria Island, Nunavut	August 1905	+4
King Point, Yukon	December 1905	−28
Nome, Alaska	August 1906	+7

a What is the difference between summer and winter temperatures on King William Island? **Show** your working.

b Find which is colder, King Point or King William Island (winter) and by how many degrees. **Show** your working.

c Find the largest temperature difference between two points on the trip. **Show** your working.

d What special equipment or precautions do you think Amundsen took in order to be prepared for such a large temperature swing? **Explain**, using at least two examples.

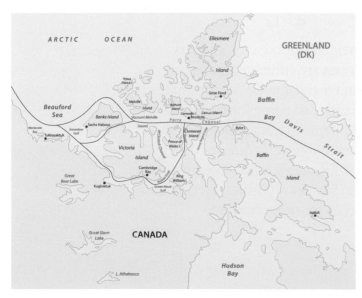

11 Evaluate these calculations that involve multiple operations.

a $-5+2(-3)$

b $10 \div -2 - (-12) \div 6$

c $(-3)^2 - (-18) \div (-3)$

d $-24 \div (-2)^3 - (-8)$

e $-16 \div (-8) + (-5)(-3)$

f $\dfrac{-10 + (-3)(6)}{2 + 3(4)}$

g $-6 - 4 - 2(-3)^2$

h $\dfrac{-2(5^2)}{-5 - (-3) + (-2)(4)}$

i $\dfrac{-6(4) + (-18) \div (+2) - 12}{2 - 9(3) + 10}$

j $(-4)^2 - (-2)^3 + (-10)(3)$

k $[(-5)(-12)] \div (+6) + (-5)$

12 By the 18th century, there were at least 35 different temperature scales in existence across the planet. For explorers, that potentially meant having to convert from one system to another, just to make sure they had the correct gear. In order to convert a temperature from degrees Fahrenheit (°F) to degrees Celsius (°C), you use the following formula.

$$C = 5(F - 32) \div 9$$

Find the Celsius temperature that is equivalent to each of the following Fahrenheit temperatures.

a 50°F **b** 5°F **c** −4°F **d** −13°F **e** −40°F

13 Create an expression using integers and at least four different operations from BEDMAS (brackets, exponents, division/multiplication, addition/subtraction) where the answer is:

a 0 **b** 3 **c** −2 **d** −12

14 If you want to explore your neighborhood, city, country or even the entire planet, you can try either geocache or EarthCache. On a device, you receive the GPS coordinates of a location where someone has hidden some form of 'treasure' which could be anything from items to trade, a logbook or information about the unique geological features of the site. If you visit a large number of EarthCaches, you can earn awards for passing certain milestones (50 sites, 100 sites, etc.). You can even create your own caches for others to find.

Suppose you visited EarthCaches at the following locations.

Location	Latitude (degrees)	Longitude (degrees)
Blue Hole, Belize	+17	−88
Stonehenge, England	+51	−2
Angkor Wat, Cambodia	+13	+104
Victoria Falls, Zambia	−18	+26
Chichen Itza, Mexico	+21	−89
Machu Picchu, Peru	−13	−73
Petra, Jordan	+30	+35
Great Barrier Reef, Australia	−18	+148
Taj Mahal, India	+27	+78

a **Write down** an itinerary to visit these destinations in order of their latitudes. **Show** your itinerary on a number line.

b Find the largest degree difference in latitude. **Show** your working.

c Find the largest degree difference in longitude. **Show** your working.

d If you traveled from Victoria Falls to the Blue Hole, find your change in latitude and longitude in degrees.

e Suppose you traveled from location to location in the order given in the table. Show that the difference (in degrees) in latitude and longitude from the Blue Hole to the Taj Mahal is equivalent to the sum of the differences between each location. **Show** your working.

15 a $\dfrac{(-75)-(-40)\div(-4)+(-5)}{[(-45)+90]\div(-5)}$

b $\dfrac{81\div(-9)+30-(-19)}{(-6)\times(-5)+(-10)}+\dfrac{100+(-60)\div(-12)-(-5)}{(80+10)\div(-9)}$

c $\dfrac{(-8)(+11)+(-6)(+7)}{(-39)-(-52)}-\dfrac{(-12)(5)+(-6)}{(+7)(-3)-(-10)}$

Summative assessment

In this summative task you will follow the voyage of a famous explorer and then plot out your own voyage of discovery. You will create a map of each trip as well as writing journal entries about life during the voyage.

Famous expedition

1   Select your explorer and voyage. You need to choose a famous explorer (e.g. Dias, Da Gama, Columbus, Drake, Vancouver, Cook) and one of their voyages. You may not use any of the voyages seen so far in this unit, but you may use a different one by the same explorer.

2   Create a map of the voyage using the system of latitude and longitude to trace the journey. On your map, you must indicate the latitude and longitude of the starting and ending positions (using positive numbers for North and East, and negative numbers for South and West). You will also indicate important locations that were visited. Round all quantities to the nearest degree.

3   Write a minimum of six short (half-page) journal entries representing what you think life must have been like on the voyage. You must include the following in your journal, written in a way that seems natural (as opposed to contrived):

- the name of the ship and a description of how many crew are on board, or of the team and its equipment (first journal entry)

- the goal of the journey and what is expected to be found

- changes in both latitude and longitude with each stop, including calculations (there must be a minimum of four stops). These must fit naturally in the journal entry.

Your final journal entry must include an expression that shows your starting latitude and longitude as well as all of the changes calculated for each. This should clearly show your ending latitude and longitude.

For example, latitude: $-75° + 23° + 5° - 8° - 15° = -70°$

Your own expedition

The next great unknown is outer space. While humans have traveled to the Moon, we have yet to explore much further than that. In this section, you will create an expedition to one or more locations in outer space.

1   Choose your location(s) and create a map of the voyage. You will need to create a grid system so that others can easily replicate your trajectory. Be mindful of the distances you need to cover when selecting your scale. Define where 'zero' is and explain your choice. Round quantities to the nearest unit.

2   Write six short (half-page) journal entries representing life on your voyage. You must include the following in your journal, written in a way that seems natural (as opposed to contrived):

   • the name of the ship and a description of how many crew are on board (first journal entry)

   • the goal of the voyage and what is expected to be found

   • at least four examples of the multiplication and/or division of positive and negative numbers

You may base your examples on the types of calculations done in this unit (e.g. force of gravity) or on others that you have researched. Your examples must include the operation as well as a worked-out solution, written in a way that fits naturally in the journal entry.

4 Algebraic expressions and equations

When studying algebra, a lot of students ask, "when am I ever going to use this in real life?" Your study of algebra in this unit will take you on a journey through puzzles and tricks that can be better understood through an algebraic analysis. However, your introduction to the rules and principles of algebra could take you on a very different journey.

 ## Identities and relationships

Wants and needs

As you grow older, you will realize that there are things that you need and things that you want, and they may not always be the same. For example, you need food to live, but you want it to be ice cream. You do not need ice cream to live! The study of wants and needs is an ideal place to apply algebraic thinking.

Electricity was once considered a luxury but it has become thought of as a basic human need. Could you live without it? Could you live without your cellphone? Have these become needs or are they still wants?

The principles behind electricity and electronic devices also use algebraic thinking, from the different ways in which circuits can be set up to the decision of which payment plan to buy.

An understanding of algebra could help you to analyse and differentiate the aspects of life that you want from those that you truly need.

At what point does a desired product become a necessity for all people? Is safety a need or a want?

Personal and cultural expression

Histories of ideas, fields and disciplines

While understanding the rules of algebra is important, placing them in the context in which they were developed can lead to a deeper understanding of why they have become such a necessary part of human existence.

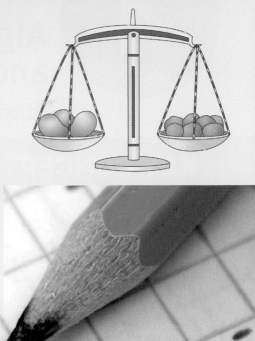

The Greek mathematician Diophantus is often referred to as "the father of algebra". However, unlike the algebra that you study, Diophantus represented his equations with words. Although he lacked the ability to use symbols, he developed a way of notating commonly occuring operations and even exponents. His ideas are often used as an introduction to algebra, solving balance problems and using manipulatives. The initial goal is to develop the concept of equality, and then develop notation and rules later.

A Persian mathematician, Muhammad al-Khwarizmi, is recognized as having established the discipline of *al-jabr*, from which the name *algebra* is derived. Al-Khwarizmi wrote several texts which are considered to be the foundations of modern algebra. While he focussed on solving equations, he was also the first to introduce the concepts of "balance", and "reduction", fundamental principles in both algebraic manipulation and equation solving. An understanding of his theories and how he developed them will clarify the rules so often associated with algebra.

4 Algebraic expressions and equations
Puzzles and tricks

KEY CONCEPT: FORM

Related concepts: Simplification and Equivalence

Global context:

In this unit, you will discover how logic and algebra have been used to create and solve puzzles and tricks. As part of the global content **scientific and technical innovation** you will see how many tricks are based on mathematical principles that anybody can apply. By learning about algebra as simply a way of expressing quantities and solving problems, you will be able to create your own puzzles and tricks and call yourself a 'mathemagician'.

Statement of Inquiry:

Producing equivalent forms through simplification can help to clarify, solve and create puzzles and tricks.

Objectives

- Defining polynomials based on the number of terms
- Writing and simplifying algebraic expressions
- Writing and solving algebraic equations and inequalities
- Applying mathematical strategies to solve problems involving algebraic equations
- Representing inequalities on a number line

Inquiry questions

F What does it mean to 'simplify'?

C What does it mean to 'be equivalent'?
How does simplification produce equivalent forms?

D Does every puzzle have a solution? Explain.
Can every trick be explained? Explain.

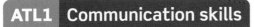

ATL1 Communication skills

Make inferences and draw conclusions

ATL2 Creative-thinking skills

Apply existing knowledge to generate new ideas, products or processes

You should already know how to:

- perform operations with integers
- perform operations with fractions

Introducing algebraic expressions and equations

For her school's Talent Night, Jeremae decided to perform a few tricks. For one of them, she projected a large monthly calendar on a screen for all to see. She then asked a volunteer to come onstage and choose a 3 by 3 section from the calendar and to sum all 9 numbers in the section. She offered him a calculator just in case! She also told him to focus on the grid he had selected so she could read his mind.

Jeremae then asked the volunteer to tell her the total and, within a matter of seconds, she told him which 3 by 3 grid he had selected.

How is that even possible? Had she really read his mind? Can you figure out the trick?

Many puzzles and tricks, including some incredible card tricks, are based on very simple mathematics.

Sun	Mon	Tue	Wed	Thu	Fri	Sat
		1	2	3	4	5
6	7	8	9	10	11	12
13	14	15	16	17	18	19
20	21	22	23	24	25	26
27	28	29	30	31		

In this unit, you will explore the 'key' to many of these tricks. However, be warned, you may never look at another magic trick or puzzle in the same way again!

Algebraic expressions

Describing expressions

An *algebraic expression* is a mathematical phrase that can contain numbers, variables and mathematical operations (addition, subtraction, multiplication and division). Algebraic expressions are often called *polynomials*, with 'poly' meaning 'many', and 'nomials' meaning 'terms'.

A *term* is a single expression, which may contain multiplication or division.

For example, $8x$, $-10xy$, $4x^2z$ and -17 are all referred to as terms.

$$\underbrace{6x^2y}_{\text{term}} + \underbrace{14y^3}_{\text{term}} - \underbrace{7x}_{\text{term}} + \underbrace{15}_{\text{term}}$$

A polynomial:

The coefficient of the first term is 6.

The coefficient of the second term is 14.

The coefficient of the third term is 7.

The constant term (the term without any variable) is 15.

Polynomials can be defined in a variety of ways. Two of those are the focus of the following investigation.

Investigation 1 – Some terminology

1 Look at the table below and use the examples to figure out the meaning of the words 'monomial', 'binomial', 'trinomial' and 'degree'.

Expression	Classification	Degree
$2y$	monomial	1
$8x + 9w$	binomial	1
$-12mn$	monomial	2
$x^2 - 4x + 2$	trinomial	2
$5z^6 - 4z^3 + 2z^2 - 1$	polynomial	6
$-7x^2y^3 + 8xy^4 - 9$	trinomial	5
$36x^2y^2z^3 - 14x^8y$	binomial	9
$76m^2n^3p^4q$	monomial	10
$2 + 8m - 4m^3 + 7m^2 - 9m^5$	polynomial	5
$-9xyz$	monomial	3
$\dfrac{2t}{7}$	monomial	1
14	monomial	0
$-17g^4 - 8$	binomial	4
$b^2 - 5b^3 + \pi$	trinomial	3
$\dfrac{4x - 9}{8}$	binomial	1
$\dfrac{5x^2 - x^3 + 8x - 1}{3}$	polynomial	3
$\dfrac{-7x^2y^5 + 8xy + 4y}{6}$	trinomial	7

▶ Continued on next page

2 Based on your results in step **1**, classify each of the following polynomials and its degree.

Expression	Classification	Degree
$17x^4 - 12x^3 + 5x$		
$\dfrac{-17y^3z}{5}$		
$14x^5 - 3\sqrt{x} + 10$		
$-8mn^2 + 2n^5$		
$9m^2n^3p^4$		

ATL1

3 Write down a definition of each of the following terms. Compare with a peer, then as a group of four. Finally, compare definitions as a class.

- monomial

- binomial

- trinomial

- polynomial

- degree of a polynomial

Reflect and discuss 1

- Explain why the words *monomial*, *binomial* and *trinomial* make sense when classifying polynomials.

- When would you use the term *polynomial* to describe an expression? Explain.

- 'The degree of a polynomial is the largest exponent that you see.' Do you agree or disagree with this statement? Explain.

The *standard form* of a polynomial is when it is written with the highest degree first and the remaining terms written in descending order of degree.

The polynomial $-3 + 3x^2 - 5x + 9x^3$ in standard form would be written as $9x^3 + 3x^2 - 5x - 3$.

When polynomials contain only one variable, they are sometimes given a specific name.

Example	Degree	Name
-4	0	constant
$6x$	1	linear
$3x^2 - 5x + 1$	2	quadratic
$-2x^3 + 6x^2 - 7x + 4$	3	cubic
$14x^4 - 5x^2 + 2x - 1$	4	quartic
$-6x^5 - 8x$	5	quintic

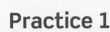

Practice 1

1 Copy and fill in the following table.

Expression	Name	Degree
$-18x^3 y$		
$-9x^2 + 2x - 21$		
$-3y^5 z^6 + 9yz^{11}$		
$11m^4 n^4 - 7k^8$		
$\dfrac{5x - 2}{3}$		
$12y^3 + 4y + 8$		
	binomial	5
	polynomial	1

2 Write the following expressions in standard form and name each based on its degree.

a $8x - 4x^3 + 7$ **b** $-2x^2 + x^4 - 11 + 7x^3 - x$

c $x - 1 - x^2$ **d** $-3x^4 - 4x^3 + 2x^5$

e $12 - x^2$ **f** $x - 11$

g $-2x^3 + 4 - 8x + 3x^2$ **h** -8

3 Write down an expression that meets the following characteristics:

a a quintic that is a trinomial   **b** a quadratic that is a monomial

c a polynomial that is a quartic   **d** a binomial that is linear

e a quadratic with no constant term   **f** a trinomial of degree 6

▶ Continued on next page

4 Solve this crossword puzzle using the clues below.

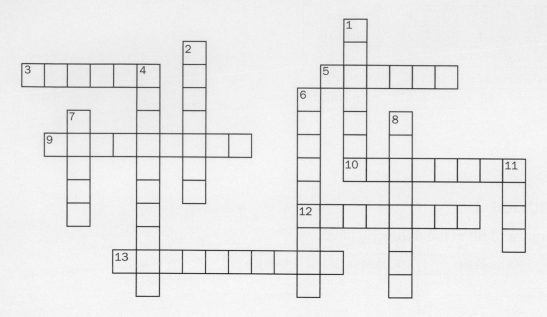

ACROSS

3   The largest exponent in a single term in a polynomial

5   Of degree one

9   Of degree two

10 Of degree zero

12 A single term

13 Many terms

DOWN

1   Of degree 5

2   Of degree 4

4   Mathematical phrase containing numbers, variables and mathematical operations

6   Three terms

7   Of degree three

8   Two terms

11 An expression containing only multiplication and division

Simplifying expressions

In the introduction to this unit, how did Jeremae know which 3 by 3 section her volunteer had chosen simply by the sum of the 9 squares?

Activity 1 – Calendar tricks

Pairs

Suppose the volunteer had chosen the following 3 by 3 section:

Sun	Mon	Tue	Wed	Thu	Fri	Sat
		1	2	3	4	5
6	7	8	9	10	11	12
13	14	15	16	17	18	19
20	21	22	23	24	25	26
27	28	29	30	31		

1 Within the 3 by 3 section, how do the numbers in each row relate to the first number in that row?

2 Within the 3 by 3 section, how do the numbers in each column relate to the first number in that column?

3 Does this apply to *any* 3 by 3 section? Explain.

4 What is the sum of the nine numbers in the highlighted section? How does this number relate to the middle number?

5 Select a different 3 by 3 section and see if the same rule applies.

6 What is the 'trick' behind this magic trick? How can Jeremae quickly figure out the total of the nine dates? Explain.

Algebra can help in understanding why this trick works. Any 3 by 3 section of the calendar looks like the following:

n	$n+1$	$n+2$
$n+7$	$n+8$	$n+9$
$n+14$	$n+15$	$n+16$

The number n is the lowest date in the grid. The date to the right of it is always one more, and the one after that is always two more. Being one week later, each date in the second row is 7 more than the one above it. Each date in the third row is 14 more than the corresponding one in the first row. But how do you add these expressions since there are variables and numbers?

$$n + n + 1 + n + 2 + n + 7 + n + 8 + n + 9 + n + 14 + n + 15 + n + 16$$

In algebra, this is called *simplifying* an expression.

 ## Investigation 2 – Like terms

1 First, it is important to be able to define what *like terms* are. Look at the following examples and determine the meaning of like terms.

Like terms	*Not* like terms
$3x$ and $-6x$	$7y$ and $9x$
$2x^2y$ and $7x^2y$	x^3y and xy^3
-13 and -10	5 and $2i$
$8mn$ and $-12mn$	$-3Mn$ and $6mn$
$4y^3$ and $9y^3$	$8z^2$ and $-11z^3$
$-2g^2p^3$, $5g^2p^3$ and g^2p^3	$3tw^2$, $-2t^2w$ and $7tw$

 2 Write down a definition of *like terms*. Compare with a peer, then as a group of four. Finally, compare definitions as a class.

3 Use the following examples to generalize the process for simplifying algebraic expressions.

Expression	Simplified expression
$3x - 6x$	$-3x$
$2x^2y + 7x^2y$	$9x^2y$
$2x - 13 - 5x - 10$	$-3x - 23$
$8mn + 7 - 12mn - 2$	$-4mn + 5$
$4y^3 + 8y^2 - 4y + 2 + 9y^3 - y + 6 - 5y^2$	$13y^3 + 3y^2 - 5y + 8$
$-2g^2p^3 + 8pg^2 + pg - 6pg - 6 + 5g^2p^3 + g^2p^3 - 10pg^2$	$4g^2p^3 - 2pg^2 - 5pg - 6$

 4 Write down the process for simplifying algebraic expressions. Be sure to use appropriate math terminology.

Reflect and discuss 2

- Why does simplifying algebraic expressions feel like simply adding and subtracting integers? Explain.

- What happens to the variables in like terms when they are added or subtracted? Explain.

Example 1

Q Simplify the following expressions.

a $n + n + 1 + n + 2 + n + 7 + n + 8 + n + 9 + n + 14 + n + 15 + n + 16$

b $6x + 5x^2 - 2 - 8x + x^2 + 10$

A **a** $n + n + 1 + n + 2 + n + 7 + n + 8 + n +$
$9 + n + 14 + n + 15 + n + 16$

> Determine which terms are like terms (highlighted).

$= 9n + 72$

> Combine like terms by adding/subtracting their coefficients.

b $6x + 5x^2 - 2 - 8x + x^2 + 10$

> Determine which terms are like terms (highlighted).

$6x^2 - 2x + 8$

> Combine like terms by adding/subtracting their coefficients. Be sure to write your answer in standard form.

Practice 2

1 For each of the groups listed, identify which term is *not* a like term, and justify your choice.

a $7x$, $2x$, $2x^2$, $-3x$ **b** $2y$, $6x$, $9y$, $-y$

c $8z$, 8, $4z$, $-8z$ **d** $-13xy$, $4xy$, $2xyz$, $-8xy$

2 Simplify each expression. Write your answer in standard form.

a $7x - 2y + x - 6y$

b $4t^3 - 6r + 9r - 10t^3 + 7 - t^3$

c $5t - 6k + 11k - 16t$

d $-9x^2y^2 + 2x^2y - 6xy^2 + 15x^2y - 12x^2y^2 + 7xy^2$

e $5x + 4xy + 3 - 9x + 7xy - 12$

f $6 - 3p + 7q - 2pq - 11q + 3p - 10$

g $-5ab + 2ab + 3a - 8b - 7ab + 13a$

h $x^2 + 3x - 13 - 7x^2 - 8$

i $-4k^2 - 9k + 3k + k^2 + 17$

j $4a^2 - 2b - 19a - 7b^2 - 10$

k $-8pq^2 + 2pq + 5p^2 - 7p^2q - 3pq^2 + 4pq$

l $y^2 - 5y^3 + 7y^2 - 8y^3 - 6y^2 + 3y^3 - 2y^2 + 10y^3$

▶ Continued on next page

3 The following is what's known as a 'magic square'.

8	1	6
3	5	7
4	9	2

a Find the sum of the numbers in each row, column and diagonal.

b How does the sum relate to the number in the middle?

c How does the sum of *all* of the numbers relate to the number in the middle? Where have you seen this result before in this unit? Does this mean a calendar is a magic square?

4 a Fill in the blanks:

2		
	5	
	1	

	7	
	5	
8		

b Now create a magic square of your own (not shown here) where every row, column and diagonal adds to 15.

5 You can even create magic squares that include algebraic expressions.

a Show that each square is a magic square by showing that the rows, columns and diagonals all have the same sum.

$2x - 2y$	$-x + 3y$	$2x - y$
$x + y$	x	$x - y$
y	$3x - 3y$	$2y$

$3y - 2x$	$-x - 5y$	$6x + 2y$
$9x - y$	x	$y - 7x$
$-4x - 2y$	$3x + 5y$	$4x - 3y$

b Create a magic square that contains numbers by substituting $x = 1$ and $y = 2$ in the magic squares above. Verify that they are magic squares.

6 Find the missing terms in each of the following magic squares.

a

	x	
	$-2x + 2y$	$4x + y$

b

		$5y$
$-2x + 3y$	x	

In Jeremae's trick, you may have noticed that the sum of the volunteer's 3 by 3 section is 9 times the middle number in that section.

n	$n+1$	$n+2$
$n+7$	$n+8$	$n+9$
$n+14$	$n+15$	$n+16$

But how do you multiply $9(n+8)$? When simplifying algebraic expressions, you may need to use the distributive property of multiplication over addition. That means that you have to multiply all the terms inside the brackets by the term outside.

Example 2

 Simplify:

a $9(n+8)$

b $4(2-x)-2(x-6)$

A   **a** $9 \times n = 9n$

       $9 \times +8 = +72$

       So, $9(n+8) = 9n+72$

> The distributive property of multiplication over addition requires you to multiply the 9 by everything inside the brackets.

> Write as one expression in standard form.

   b $4(2-x)-2(x-6) = 8-4x-2x+12$

> The distributive property of multiplication over addition requires you to multiply everything inside the brackets by the number or term in front of them.

      $= -6x+20$

> Combine like terms and write in standard form.

Pairs

Activity 2 – Pick a number

Common math tricks or puzzles often involve having someone pick a number and then perform several calculations with that number. The mathemagician then either tells the person the final result or their original number. However, these tricks are actually based on simplifying algebraic expressions.

1 Try this one.

- Write down a number.

- Add 5 to it.

▶ Continued on next page

- Double the result.

- Subtract 8.

- Triple the result.

- Add 12

- Divide the new number by 6

- Now take away the original number.

Is the answer 3?

2 Call the original number 'n'. Follow the instructions, applying them to your expression. (You may need to use brackets in some steps.) Show that the result is always 3.

3 Justify why the following trick works using algebra. You will need to use two variables this time.

- Pick a number from 1 to 9 and double it.
  Add six to this result.
  Finally, multiply by 5.

- Pick another number from 1 to 9 and add it to the last number you got.
  Now subtract 30.

- The digit in the tens position is the original number and the digit in the units position is the second number chosen!

ATL2 4 Explain the trick used to get the original number to be in the tens position.

5 How would you get a number to be in the hundreds position? Explain.

Practice 3

1 Simplify the following expressions. Write your answers in standard form.

a $2(x - 12)$

b $5(9a - 6b)$

c $-2(-3g + 3h)$

d $6(2n + 3) + 8n - 1$

e $5(2w - 9) - (3w + 7)$

f $-1(8 - 3x)$

g $4(k - 6) - (10 - 2k)$

h $3s - 2(4s - 9)$

i $8(2y + 7) + 4(2 - 4y)$

j $3(4X + 6Y - 7) + 5(2Y - 9X + 6)$

k $-2(3a - 6v) - (5a + 8v + 7)$

l $2(7x^2 + 8x - 2) - 6(2x - 3x^3 + 2x^2)$

m $3(9i + j + 3) - 4(10i + 3j + 12)$

n $-(9x^2y + 8 - 3x^2y^2 + 13x) - 2(3x^2y^2 - 12 + 16x + 16x^2y)$

▶ Continued on next page

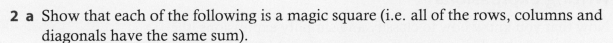

2 a Show that each of the following is a magic square (i.e. all of the rows, columns and diagonals have the same sum).

a

$2(a + b)$	$3a + b$	$a - 12b$
$a - 17b$	$2a - 3b$	$3a + 11b$
$-3(-a - 2b)$	$a - 7b$	$2(a - 4b)$

b

$3(a - b)$	$-9a + 12b$	$3b$
$-5(a - 2b)$	$2(-a + 2b)$	$-(2b - a)$
$-4a + 5b$	$5a - 4b$	$-7a + 11b$

c

$-23a + 16b$	$5(3a - 2b)$	$-19a + 3b$
$5(-a - 2b)$	$-3(3a - b)$	$-13a + 16b$
$a + 3b$	$-33a + 16b$	$5(a - 2b)$

d

$6a$	$7a^2 - 11a + 24$	$-a^2 - 7a - 6$
$a^2 - 17a$	$2(a^2 - 2a + 3)$	$-3(-a^2 - 3a - 4)$
$5a^2 - a + 18$	$-3(a^2 + 4 - a)$	$2(2a^2 - 7a + 6)$

e Create magic squares with numbers by substituting $a = -1$ and $b = -2$ in each of the squares above. Verify that each is a magic square.

3 Fill in the boxes with algebraic expressions to create magic squares.

a

$-3m^3 + 5m^2$	$-2(m^3 - 2m^2)$	
$2(2m^3 - 3m^2)$		

b

$2(-3a + b)$		$-(-2a - 4b)$
	$-4(a + 2b)$	

c

		$-4t + 5u$
		$-2(-3t - 2u)$
		$-5(t - 3u)$

4 Use algebra to justify the results in the following tricks. Show your working. Call the original number that is chosen 'n'.

a Think of a number (n). Subtract 3, then multiply the result by 4. Add 32, then divide the result by 4. Finally take away the original number. The answer is 5.

b Think of a number. Add 21. Double this result. Subtract 2. Double this result again. Add 20. Divide this number by 4. Subtract 25. You have your original number back!

c Think of a number and add 8 to it. Multiply this result by 6. Now add 2 and then divide this result by 2. Add 2 again. Finally, divide by 3 and then subtract the original number. The answer is 9.

Writing expressions

Now that you know how to classify and simplify expressions, the next step is to be able to write your own. Being able to translate from English to mathematics is one of the most important skills in algebra, because that is how you model problems and puzzles and, eventually, solve them.

Activity 3 – Translating into algebra

1 Suppose that, during a trick, your volunteer has chosen a number. Since you don't know what it is, you will represent it as a variable. Common choices are x or n, but you may choose whatever letter you like.

Translate the following words into algebraic expressions using the variable(s) of your choice.

> twice the number

> multiply the number by 5 and add 2

> the next consecutive number

> 12 minus the number

> the number minus 12

> add 6 to the number, then divide the sum by 3

> 3 times the number plus a new number

> the number of sisters plus the number of brothers

> the sum of the next two consecutive numbers

> four more than the number of pets, minus the number of siblings

> the sum of ten and some number is subtracted from another number

> the difference between two numbers is divided by 4 then multiplied by −5

> the sum of two numbers is added to their product then halved

▶ Continued on next page

2 Translate the steps of this trick using algebra to justify its final result. This trick is best done if you can't see the volunteers' slips of paper.

- *Ask three students to take three slips of paper each and write a whole number on one side of one of them.*

- *Have the students flip that paper over and write the next consecutive number on the other side.*

- *Have them write the next consecutive number on the second slip of paper, and the next number after that on the back.*

- *Ask them to repeat the process one more time with the third slip of paper so that each student has three slips of paper with six consecutive numbers on them.*

- *Ask them to place the slips of paper on the desk with the higher number face up. As a final step, have each student flip any one of the slips of paper over, and then add up the three numbers they see.*

3 What expressions did you write down for the numbers on the six slips of paper? Compare your answers with those of a peer.

4 If any student tells you their lowest number, you can tell them the sum of the three numbers by simply multiplying that number and then adding 8. Justify why this works by looking at all of the combinations of 3 slips of paper, with one of them flipped over so that the lower number faces up.

Practice 4

1 Write a simplified algebraic expression using each written description:

a Twice a number, plus 6

b The sum of three consecutive integers

c 13 minus the number of students multiplied by 3

d The number of boys multiplied by the number of girls

e The sum of twelve times the number of pets and five times the number of pets

f Half the number of cards

g The sum of a number and 4 is multiplied by 16 then subtracted from 40

h The product of triple a number and double a different number is added to eleven then halved

▶ Continued on next page

2 Create a written description that would fit these algebraic expressions:

a $4x + 10$

b $2(x - 6)$

c $\dfrac{7x - 8y}{3}$

d $\dfrac{3x(y + 6)}{2}$

3 Use algebra to justify the results in the following tricks. Show your working.

a Think of your age in years. Multiply that number by 2 and then add 10. Multiply this result by 5. Now add the number of siblings you have. From this result, subtract 50. The first two digits of your final answer represent your age and the last digit is the number of siblings you have!

b Ask a volunteer to multiply the first number of his or her age by 2. Tell them to add 3. Now tell them to multiply this number by 5. Finally, have the person add the second number of his or her age to the figure and then tell you their answer. Deduct 15 and you will have their age.

> For this trick, you need a volunteer who is at least 11 years old.

c Write down the day of month on which you were born (just the date, not the month nor year).

Multiply that number by 20.

Double today's date and add it to your answer.

Now, multiply this number by 5.

Add the month of your birthday (e.g. March is 3, November is 11).

From this result, subtract ten times today's date.

The four-digit number you are left with represents your birthday, with the date of your birthday first and the month at the end!

4 Try this trick with someone whose birthday you don't know.

Ask him/her to write down the number of the month in which he/she was born. For effect, turn your back to the person and have them complete the following instructions as you say them.

Take the birth month, double it and then add 5. Multiply the result by 50. To this number, have the person add his/her current age. From this most recent number, ask him/her to subtract 365.

At this point, turn around and ask the person to tell you the final number he/she got. Now, in your head, add 115 to the result you just heard. The first digit of your result is the person's birth month and the remaining digits represents the person's current age!

Justify why this works using algebra.

▶ Continued on next page

Justify the tricks in questions 5 through 7 using algebra. (You may want to amaze your friends with them!)

5 Have someone choose two different numbers between 1 and 9. Ask him or her to determine the two possible two digit numbers that can be formed from the single digit numbers (e.g. 3 and 5 would make 35 and 53).

Ask him/her to add the 2 two-digit numbers together, then divide that number by the sum of the two single digit numbers.

You then proclaim, 'The answer is 11.'

6 There are people who consider some shortcuts in mathematics to be 'tricks'. For example, there are divisibility rules that make it very easy to figure out if a number is divisible by 2, 5 or 9. Why do these work?

a State the divisibility rule for 3 and 9. Explain the rule with an example.

b If a 3-digit number has the digits ABC, represent this number in expanded form.

c Show that the number represented by ABC is equal to $99A + 9B + A + B + C$. Hence, describe why the divisibility rule for 3 and 9 works.

7 Have someone pick two numbers and write them one on top of the other (in a column). Have them find the next number in the list by finding the sum of the two numbers above it. Have him/her write it down and repeat the process until there are ten numbers in the list.

Tell the person you can add those 10 numbers faster than anyone, including anyone with a calculator. Have them try to find the sum before you do, using mental math or a calculator.

(All you have to do is just multiply the 7th term by 11!)

For an easy way to multiply any number by 11, visit the site **themathworld.com** and search for 'math tricks'.

Formative assessment

'Perimeter magic triangles' are triangles that contain only consecutive numbers and in which the sum of each of the sides is always the same. 'Order 3' triangles have three values per side while 'order 4' triangles have four values per side. Perimeter magic triangles can come in any order greater than or equal to 2.

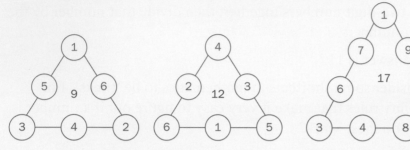

Order 3 perimeter magic triangles   Order 4 perimeter magic triangles

1 Show that the following triangles are perimeter magic triangles.

a
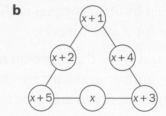

b

> In Question **1**, parts **a** and **b** are order 3 perimeter magic triangles; part **c** is order 4.

c
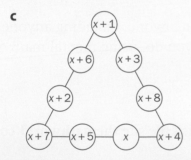

ATL2

2 Create your own perimeter magic triangles by doing the following:

a In question **1a**, replace x by $3(m + 3)$ and multiply each constant term by $(m - 2)$. For example, $x + 3$ would become $3(m + 3) + 3(m - 2)$. Show that the three sides all simplify to the same expression. Classify this expression in as many ways as possible.

b In question **1c**, replace x by $-2(b - 1)$ and multiply each constant term by $(3 - 2b)$. Show that the three sides all simplify to the same expression.

> Does every puzzle have a solution? Can all tricks be explained?

ATL2

3 Create your own magic triangle of order 5 with consecutive numbers, using algebraic expressions (as in question **1**). Show that you have created a perimeter magic triangle.

Equations and inequalities

Solving equations

An *algebraic equation* is a mathematical statement that sets two mathematical expressions equal to each other. Solving an algebraic equation is about the approach you take; you apply different operations until you have the desired result. It is not an abstract idea, but a problem-solving methodology that you use all the time, perhaps without even knowing it.

A similar methodology is used to solve balance puzzles, like those in the following activity.

Activity 4 – Balance puzzles

The following diagrams show a balance with a variety of objects on each side. 'Solve' each balance to find out how much one blue circle weighs in terms of the other shapes. Record what you do at each step.

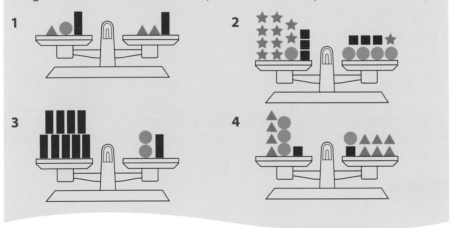

http://nlvm.usu.edu/en
On the website, work through the Algebra Balance Scales activity to solve linear equations using a balance beam representation. There is one activity using positive numbers and one that incorporates negative numbers.

Solving an equation involves the same idea: whatever operation you perform on one side of the balance/equality, you must also do on the other side. The following investigation will help you generalize a procedure to solve one- and two-step equations.

Investigation 3 – Balance puzzles

1 Look at the following balance puzzle representations. Solve the balance puzzle and perform the same operation on the equation. Write down your final answer. The first example is done for you.

Example:

Equation:

$b + 2 = 4$

Remove two weights from each side.

$b + 2 - 2 = 4 - 2$

One b weighs 2.

$b = 2$

a Balance puzzle: left: 8 weights; right: 2 bags

b Balance puzzle: left: 9 weights; right 1 bag and 4 weights

c Balance puzzle: left: 3 bags; right: 9 weights

d Balance puzzle: left: 3 bags and 1 weight; right: one bag and 7 weights

ATL1 **2** Generalize the procedure to solve a one-step equation.

3 Look at the following equations and their balance puzzle representations. Solve the balance puzzle and perform the same operation on the equation. Write down your final answer.

a

$4b = b + 6$

b

$2b + 3 = 5b$

▶ Continued on next page

c

$$3b + 4 = 6b + 1$$

d

$$7b + 3 = 8b + 1$$

e

$$7b + 2 = 4b + 11$$

ATL1 **4** Generalize a process for solving two-step equations.

Reflect and discuss 3

- What represents the 'balance' in each equation that you write? Explain.

- Looking at how you solved the equations, how do you know which operation ($+$, $-$, $\times$, $\div$) to perform? Explain.

- Looking at how you solved the balance puzzles and equations, how do you decide which side your variable will be? Explain.

- What steps would you take to solve the following equation: $3x - 5 = 7$?

Example 3

 Solve the equation $2a + 7 = 4a - 13$ for the variable a.

$$2a + 7 = 4a - 13$$
$$2a - 2a + 7 = 4a - 2a - 13$$
$$7 = 2a - 13$$
$$7 + 13 = 2a - 13 + 13$$
$$20 = 2a$$
$$\frac{20}{2} = \frac{2a}{2}$$
$$10 = a$$

Decide to have the variables on the right side of the equals sign since the coefficient of a is greater on that side ($4 > 2$). Since you use the opposite operation, subtract $2a$ from each side. Simplify.

In order to isolate the variable, add 13 to both sides of the equation.

Divide both sides of the equation by 2 to solve for a.

Go to **https://illuminations.nctm.org** and search for 'Algebra Tiles'
Under the 'Solve' tab, use the algebra tiles to help you visualize and solve algebraic equations.

Example 4

 Solve the equation $3(a - 2) = 2(a - 2)$ for the variable a.

$$3(a - 2) = 2(a - 2)$$

$$3a - 6 = 2a - 4$$

$$3a - 6 - 2a = 2a - 4 - 2a$$

$$a - 6 = -4$$

$$a - 6 + 6 = -4 + 6$$

$$a = 2$$

Before making decisions about where you would like the variables to be, you need to remove the brackets.

Use the distributive property of multiplication to remove the brackets.

Decide to have the variables on the left side of the equals sign since the coefficient of a is greater on that side ($3 > 2$). Therefore, subtract $2a$ from each side.

Simplify.

Add 6 to each side to isolate the variable.

Simplify.

Practice 5

1 Write down an equation representing each balance puzzle. Solve both the balance puzzle and the equation. Write down all of your steps.

a

b

c

d

2 Solve for the variable in the following equations.

 a $10 = x + 5$ **b** $2p = -6$ **c** $3z + 1 = 2z$ **d** $4s = 7s + 6$

 e $4m + 6 = 6m - 4$ **f** $3g - 5 = 2g - 8$ **g** $4c + 7 = 2c - 1$ **h** $\dfrac{x}{5} = 4$

 i $-x + 4 = 2x + 1$ **j** $4(x + 2) = -2(x - 1)$ **k** $3x = 4(x - 9)$ **l** $1 + 8w = 10 - w$

 m $\dfrac{y}{4} + 7 = 2$ **n** $\dfrac{a}{3} = a - 6$ **o** $\dfrac{2t}{7} - 10 = -2t + 4$ **p** $\dfrac{3d}{4} + 4 = d + 2$

3 Dmitry, Darius and Diana solved the following equation: $2m - 5 = 4m + 1$

 a Dmitry began solving the equation by adding 5 to each side.
  Solve the equation using Dmitry's method.

 b Darius began solving the equation by subtracting $2m$ from each side.
  Solve the equation using Darius' method.

 c Diana began solving the equation by subtracting $4m$ from each side.
  Solve the equation using Diana's method.

 d Which method(s) could be considered 'correct'? Which method is easiest?
  Explain your answers.

Activity 5 – Creating puzzling equations

You can create your own 'puzzling' equations by starting off with the answer you want and then performing the same operation to both sides of the equation. For example:

$Q = 2$	Start off with the answer.
$4Q = 8$	Multiply both sides by a number, such as 4.
$4Q + Q = 8 + Q$	Add the same amount to each side, such as Q.
$5Q = Q + 8$	Simplify and write in standard form (or not!).
$5Q - 3 = Q + 8 - 3$	Subtract the same amount from each side, such as 3.
$5Q - 3 = Q + 5$	Simplify.

1 Create three complicated equations like the one above. Make sure at least one of them has a negative number as an answer.

2 Trade equations with the members of your group and solve each other's.

3 Select the one equation that you would like to present to the class that you think will challenge your peers!

Writing equations

Number puzzles don't always have a surprise ending, like many tricks do. Sometimes, a puzzle is about finding whether a solution is even possible. For example, can you find three consecutive numbers that, when added together, equal the sum of the next two consecutive numbers? Where would you even begin to solve this puzzle?

Example 5

Q Find three consecutive numbers that, when added together, equal the sum of the next two consecutive numbers.

A Let n = the original number.

> Define your first variable.

Let $n + 1$ = the first consecutive number.

Let $n + 2$ = the second consecutive number.

> Since all of the numbers relate to one another, write them in terms of the one variable n.

Let $n + 3$ = the third consecutive number.

Let $n + 4$ = the fourth consecutive number.

$n + (n + 1) + (n + 2) = (n + 3) + (n + 4)$

> Write an equation that represents the given situation. In this case, adding the first three consecutive numbers is equal to adding the next two.

$3n + 3 = 2n + 7$

> Simplify both sides.

$3n + 3 - 2n = 2n + 7 - 2n$

> Perform the opposite operation so that the variable is on only one side of the equation.

$n + 3 = 7$

> Simplify.

$n + 3 - 3 = 7 - 3$

> Isolate the variable.

$n = 4$

Therefore, $n + 1 = 5$, $n + 2 = 6$, $n + 3 = 7$ and $n + 4 = 8$

> Find the other values.

$4 + 5 + 6 = 7 + 8$

> Verify your answer.

The consecutive numbers are 4, 5 and 6.

> Answer the question explicitly.

Practice 6

1 A classmate has developed a few tricks of her own. Represent each of the following with an equation and solve it to justify the solution to her tricks.

a She says, 'Choose four consecutive integers. If you tell me their sum, I can tell you the numbers you chose.' You tell her that their sum is 102. What numbers does she say you chose?

b 'Think of a positive number. Multiply it by 4 and then add 3. Double your original number and add it to this result. Now subtract 9. Finally, divide this result by 6. If

▶ Continued on next page

you tell me the result, I can tell you what number you started with.' You tell her the result is 40. What number does she say you originally chose?

c 'Think of three consecutive positive numbers. Double the smallest one and subtract the largest one from that result. If you tell me the result, I'll tell you the numbers you started with.' You tell her the result is 70. What numbers does she say you originally chose?

2 a Hector tells Amaya, 'I know three consecutive numbers such that, if you add them, the result is one more than double the next consecutive integer.' Find the three consecutive numbers that Hector is referring to.

b Amaya responds with, 'Well, I know two numbers that differ by 6 such that, when you add them, you get 5 times the smaller one.' Find the numbers Amaya is referring to.

3 Luan thinks equations are like little puzzles. He created his own number puzzle and told his friend, 'I am thinking of an integer. If I multiply my number by 2 and subtract 4, I get the same result as when I multiply it by 3 and add 5. What is my number?' Write down an equation that represents Luan's puzzle and find his number.

4 Create your own number puzzles that lead to an equation. Trade with a peer and solve his/her puzzles.

Equations in cryptology

Cryptology is the science of secrets and secure communication. It covers both cryptography (the process of writing various ciphers to keep messages secret) and cryptanalysis (the science of finding weaknesses in these ciphers). It is essentially the art of creating puzzles that are very hard to solve. The harder they are to solve, the more secure the communication. To understand how ciphers work, you are going to look at how Julius Caesar used them to communicate with his armies.

Did you know?

Julius Caesar (100 BC to 44 BC) was a Roman politician and army general who played a significant role in the rise of the Roman Empire. As emperor, he controlled vast territories in Europe, Africa and Asia and ruled one of the most powerful empires in history. During his military career, Caesar used a basic substitution cipher to send messages to his soldiers that could not be read easily if intercepted by his enemies. Caesar used a wheel to help him encrypt and decrypt messages quickly.

Caesar's cipher works by replacing each letter in the alphabet by another letter that is a fixed number of positions away in the alphabet. For example, with a right shift of 3, the letter 'a' would be replaced with the letter 'D'; the letter 'm' would be replaced with the letter 'P', etc. Caesar often used the shift of 3 units, but you could shift the letters by any number. This is called the *key,* and as long as both the sender and receiver know the key, coded messages can be sent and then decoded by the receiver.

Below is the typical shift of 3 that Caesar commonly used. The top line is the *plaintext* of the message and the bottom line is the *ciphertext* or *coded text*.

a	b	c	d	e	f	g	h	i	j	k	l	m	n	o	p	q	r	s	t	u	v	w	x	y	z
D	E	F	G	H	I	J	K	L	M	N	O	P	Q	R	S	T	U	V	W	X	Y	Z	A	B	C

WEB LINK

Caesar used a wheel to help him encrypt and decrypt messages easily. Search for 'Caesar cipher disc' on the **raft.net** website for instructions on how to create your own Caesar cipher disc.

Activity 6 – Caesar cipher

1 Decode the following famous quotes made by Caesar using this cipher with a key of 3.

   L FDPH, L VDZ, L FRQTXHUHG

   OHW WKH GLH EH FDVW

ATL2

2 Clearly, cracking the code would be a lot more difficult if you did not know the key. A different key has been used to encode the message below. Try to decode it with a partner and write down any strategies you used so then you can share with the class to get a full list of working strategies to use.

   NGYNARNWLN RB CQN CNJLQNA
   XO JUU CQRWPB

3 Describe the shift that was used to encode the message. (e.g. right shift of 4, left shift of 10, etc.)

Reflect and discuss 4

- How many possible keys are there for Caesar ciphers?
- What are the strengths of the Caesar cipher (regardless of the shift)?
- What are the weaknesses of the Caesar cipher (regardless of the shift)?
- Why do you think these ciphers were effective?
- Is there a shift you can do so that the same algorithm (mathematical procedure) can be used for both encoding and decoding? Why does it work?

Substitution ciphers can be written as algebraic equations to help describe codes more efficiently.

First, every letter in the alphabet is assigned a number.

A	B	C	D	E	F	G	H	I	J	K	L	M
1	2	3	4	5	6	7	8	9	10	11	12	13

N	O	P	Q	R	S	T	U	V	W	X	Y	Z
14	15	16	17	18	19	20	21	22	23	24	25	26

Now, just like with the wheel, there can be no particular finishing position; once you go through the alphabet then it just starts all over again. This concept is called modular arithmetic. As with a traditional clock, we count from 1 to 12, then start again at 1 instead of going to 13. So with numbers representing the alphabet, the number 27 would be represented as a 1 again. This is denoted as (mod 26).

Example 6

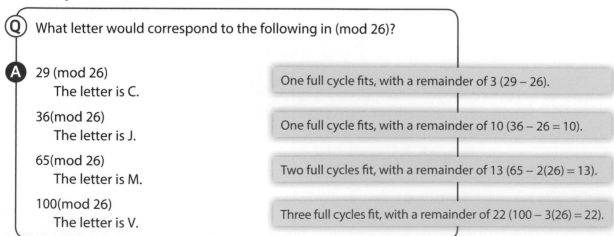

Q What letter would correspond to the following in (mod 26)?

A 29 (mod 26)
   The letter is C.

One full cycle fits, with a remainder of 3 (29 − 26).

36 (mod 26)
   The letter is J.

One full cycle fits, with a remainder of 10 (36 − 26 = 10).

65 (mod 26)
   The letter is M.

Two full cycles fit, with a remainder of 13 (65 − 2(26) = 13).

100 (mod 26)
   The letter is V.

Three full cycles fit, with a remainder of 22 (100 − 3(26) = 22).

Cryptography often uses a notation and process that resembles solving equations.

Suppose the plaintext is the letter x, the cipher (coded) text is the letter y and the key is 'k'. Then the encryption rule or function can be written as:

$$y = x + k$$

To decrypt the message you need to undo the code, something called 'finding the inverse'. Finding the inverse is like solving an equation and it requires using the opposite operation. To find the inverse, swap the position of the x and y, and solve for y (called 'isolating y') to find the decryption function:

If $\quad y = x + k$	
$x = y + k$	Swap the x and the y.
$x - k = y + k - k$	Solve for y.
$x - k = y$	

Just like solving an algebraic equation, you do the opposite operation to find the solution.

Example 7

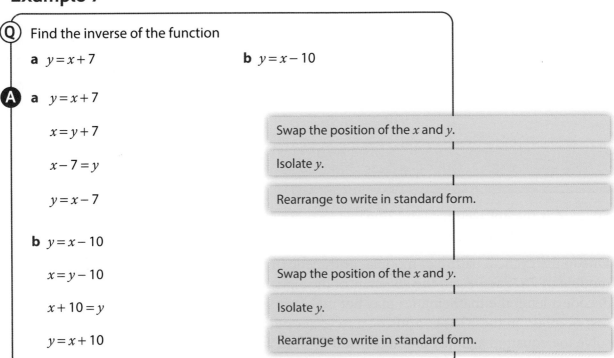

Q Find the inverse of the function

a $y = x + 7$ **b** $y = x - 10$

A **a** $y = x + 7$

    $x = y + 7$ Swap the position of the x and y.

    $x - 7 = y$ Isolate y.

    $y = x - 7$ Rearrange to write in standard form.

  b $y = x - 10$

    $x = y - 10$ Swap the position of the x and y.

    $x + 10 = y$ Isolate y.

    $y = x + 10$ Rearrange to write in standard form.

Activity 7 – Name encryption

Pick a value that is no more than 5 for your shift (*k*) and apply the process to your name. Convert your plain text name into a number and work through the encryption equation to find your cipher-text name. Have your name be between 5 and 10 letters, so add some letters of your last name if you have a name under 5 letters and cut it short if it is above 10 letters. Write it down on a small piece of paper supplied by your teacher. Hand it in to your teacher who will redistribute the coded names to the class. Your task is to use your powers of deduction and decoding skills to figure out the person's name and their shift key.

Much more complex ciphers are possible. These codes are considerably harder to break.

For example, sending the message 'HELLO' using encryption function $y = 2x + 10$ would mean:

H	8	$y = 2(8) + 10 = 16 + 10 = 26$		Z
E	5	$y = 2(5) + 10 = 10 + 10 = 20$		T
L	12	$y = 2(12) + 10 = 24 + 10 = 34$	$34(\text{MOD } 26) = 8$	H
L	12	$y = 2(12) + 10 = 24 + 10 = 34$	$34(\text{MOD } 26) = 8$	H
O	15	$y = 2(15) + 10 = 30 + 10 = 40$	$40(\text{MOD } 26) = 14$	N

Decrypting the message 'ZTHHN' would begin by finding the inverse function:

$$x = 2y + 10$$
$$x - 10 = 2y$$
$$\frac{x - 10}{2} = y$$
$$y = \frac{x}{2} - 5$$

So, as Z is the 26th letter of the alphabet, $\frac{26}{2} - 5 = 13 - 5 = 8$ and the 8th letter of the alphabet is H.

Due to the cyclical nature of modular arithmetic (in this case, a module of 26 to represent the 26 letters of the alphabet) when you decode a letter the mathematics may not make sense with your first number. In these cases, add 26 on to your original number and solve the equation again. Continue to add 26 until your solution is an integer between 1 and 26 so you can then decode the letter.

For example, attempting to decode the letter H:

H is the 8th letter of the alphabet.

Using the inverse function $y = \dfrac{x}{2} - 5$

gives $y = \dfrac{8}{2} - 5 = -1$

and does not relate to a letter of the alphabet.

So, instead of 8, use 8 + 26:

$$y = \frac{8 + 26}{2} - 5 = 12$$

And the 12th letter of the alphabet is L.

Decrypt the remaining letters using the same inverse function to get the rest of the word.

Did you know?

At the end of World War I, a German engineer invented the Enigma Machine, which was used extensively in World War II. A message was typed and then coded onto 3 to 5 wheels. The person receiving the message could decode it, as long as he knew the exact settings of these wheels. Alan Turing, a British mathematician, used mathematics to eventually break the Enigma machine code, which helped end World War II.

Example 8

Q Find the inverse of the function:

a $y = 2x + 14$ 　　　　　　　　**b** $y = 4x + 12$

A **a** $y = 2x + 14$

$x = 2y + 14$ 　　　　　　　Swap the position of the x and y.

$x - 14 = 2y$ 　　　　　　　Subtract 14 from both sides.

$\dfrac{x - 14}{2} = y$ 　　　　　　　Divide by 2.

$y = \dfrac{x}{2} - 7$ 　　　　　　　Rearrange and simplify.

▶ Continued on next page

b $y = 4x + 12$

$x = 4y + 12$ | Swap the position of the x and y.

$x - 12 = 4y$ | Subtract 12 from both sides.

$\dfrac{x-12}{4} = y$ | Divide by 4.

$y = \dfrac{x}{4} - 3$ | Rearrange and simplify.

Practice 7

1 Find the inverse of the following functions.

a $y = 2x - 10$ **b** $y = \dfrac{x+6}{5}$ **c** $y = -6x - 12$ **d** $y = 2 - \dfrac{1}{2}x$

e $y = -2x + 14$ **f** $\dfrac{x-2}{3} = y$ **g** $5 + \dfrac{1}{3}x = y$ **h** $y = 3 - \dfrac{x}{4}$

2 The following message was encrypted using the basic encryption function $y = 3x + 21$.

HXCSJHXCVDZ   HXBJZ   CSJ   LNWEG   PN   XWNFKG

a Find the inverse and then use it to decode the message.

b Do you agree with the decrypted statement?

Did you know?

Mary, Queen of Scots, sent coded messages to her co-conspirator Anthony Babington while she was imprisoned by Elizabeth I. It was the capture and decoding of one of these messages, in which she consented to a plan for the assassination of Elizabeth I, that led to her execution.

Reflect and discuss 5

- Who needs to keep secrets? What organizations need to keep secrets? Who would want to know these secrets?

- Was it easier to encrypt or decrypt messages?

- Do you think you could decode a message without the encryption function? Explain.

- What problems do you see with this type of encryption, in which both the sender and receiver need the key?

Representing inequalities

Not all tricks or puzzles involve knowing the final result. Some tricks start off with choosing a number less than or more than a certain amount. An *inequality* or *inequation* is a statement that compares two quantities or expressions, for example, $x \geq 7$ or $2x + 1 < 5$.

Whereas an equation is well represented with a balance, inequalities are best represented on a number line.

Investigation 4 – Representing inequalities

1 Write each of the following inequalities in words (e.g. '*y* is greater than or equal to 7'). Then match each inequality with its number line representation.

Inequality

$5 \geq m$

$p > -7$

$-2 < h$

$-5 \geq x$

$r \geq 4$

$Z \geq 7$

$K \leq 0$

$0 > f$

$3 > t$

$Q \geq -2$

$W < -5$

Representation

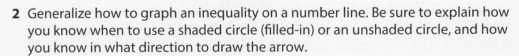

a
b
c
d
e
f
g
h
i
j
k

2 Generalize how to graph an inequality on a number line. Be sure to explain how you know when to use a shaded circle (filled-in) or an unshaded circle, and how you know in what direction to draw the arrow.

Reflect and discuss 6

- What would it look like if you graphed $x = 7$ instead of $x < 7$ or $x > 7$?
- How do you think you would graph $-2 < x \leq 4$? Explain.

Practice 8

1 Represent these inequalities on a number line.

a $d \leq -3$ **b** $m > 4$ **c** $u \geq -1$

d $z < -5$ **e** $p \geq 8$ **f** $t \leq 0$

2 Write an inequality to represent each of the following. Use the variable n.

a
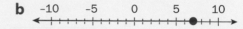

b
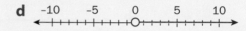

c
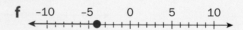

d

e

f

3 The following are words taken from the beginning of a number trick. Represent each of them on a number line.

a 'Think of a number less than 20.'

b 'Write down any number greater than 100.'

c 'Choose a number less than or equal to 50.'

d 'I'm thinking of a number greater than 10.'

Solving inequalities

Like equations, inequalities can be solved. When you have an equation, as long as you perform the same operation on both sides, the equation remains equal. What happens when you perform an operation on both sides of an inequality? Does the statement of inequality remain true? You will figure this out in the next investigation.

Investigation 5 – Operations on inequalities

Perform the following operations on each side of the original inequality and write your result in the middle column. Interpret your result in the last column.

Original inequality	Operation	Resulting inequality	Is the statement still true?
$10 > 4$	Add 2 to both sides.		
$10 > 4$	Subtract 2 from both sides.		
$10 > 4$	Multiply both sides by 2.		
$10 > 4$	Divide both sides by 2.		
$10 > 4$	Multiply both sides by -2.		
$10 > 4$	Divide both sides by -2.		

1 Under what conditions did the inequality remain true? When was it not true?

2 Verify your generalization starting with the inequality $-6 \leq 3$, using the same operations as before but with different numbers (e.g. multiply by 3).

3 Write down a true statement of inequality and verify your generalizations again by adding, subtracting, multiplying and dividing by both positive and negative numbers.

4 What could you do to make an inequality true after you have multiplied or divided by a negative number?

Reflect and discuss 7

- Why do you think an inequality remains true no matter what number you add or subtract to each side?

- Why do you think an inequality becomes false when you multiply or divide by a negative value?

- When solving an inequality, what will you have to remember to do when you multiply or divide by a negative number in order to keep the inequality true?

Solving inequations involves the same process as solving equations. The only difference is that you will have to switch the direction of the inequality whenever you multiply or divide by a negative number. The solutions to inequalities are often graphed on a number line.

Example 9

Q Solve and represent the solution on a number line.

a $2f + 4 > f - 2$

b $3(g + 2) \geq 5(g - 2)$

A **a** $2f + 4 > f - 2$

$2f + 4 - f > f - 2 - f$

| | Isolate the variable on the left side of the inequality. Subtract f from each side. |

$f + 4 > -2$

| | Simplify. |

$f + 4 - 4 > -2 - 4$

| | Isolate the variable by subtracting 4 from both sides. |

$f > -6$

| | Simplify. |

| | Represent on a number line. It is an open circle since it is just 'greater than'. |

b $3(g + 2) \geq 5(g - 2)$

$3g + 6 \geq 5g - 10$

| | Use the distributive property of multiplication to expand each side. |

$3g + 6 - 3g \geq 5g - 10 - 3g$

| | Isolate the variable on the right side of the inequality. Subtract $3g$ from each side. |

$6 \geq 2g - 10$

| | Simplify |

$6 + 10 \geq 2g - 10 + 10$

| | Isolate the variable by adding 10 to both sides. |

$16 \geq 2g$

| | Simplify. |

$\dfrac{16}{2} \geq \dfrac{2g}{2}$

| | Solve by dividing each side by 2. |

$8 \geq g$

| | Represent 'g is less than or equal to 8' on a number line. It is a closed circle since it is 'less than or equal to'. |

Reflect and discuss 8

- Part **b** of Example 9 could have been solved by subtracting 5g from each side instead, so that the variable would be isolated on the left side of the inequality. Solve the inequality using this as a first step. What do you notice about your answer compared with the one in the example?

- If the variable is on both sides of the inequality, describe a way to avoid having to divide by negative values.

- Why is it important to read the solution before representing it on a number line? Explain.

Practice 9

1 Solve each inequality. Represent each solution on a number line.

   a $x+5<9$ **b** $2t>-6$ **c** $-5m\geq-15$

   d $\dfrac{g}{2}\geq-3$ **e** $-2y-7<3$ **f** $4p-3\leq-p+7$

   g $3(w-3)\geq-9$ **h** $7\geq4w-1$ **i** $v-4<1+6v$

   j $-2(w-1)\geq3(w+4)$ **k** $7-y\geq4(3+y)$ **l** $3d-4+2d\leq-(d+16)$

 2 Create a 'puzzling inequality' by starting with each of the following inequalities and performing operations on both sides (like you did with equations). Then, share with a peer and solve their inequalities.

   a $y<-2$ **b** $6>n$ **c** $p\geq-4$ **d** $h\leq1$

Reflect and discuss 9

- How is solving an inequality similar to solving an equation? Explain, using an example.

- How is solving an inequality different than solving an equation? Explain.

Unit summary

An *algebraic expression* is a mathematical phrase that can contain numbers, variables and mathematical operations (addition, subtraction, multiplication and division). Algebraic expressions are often called *polynomials*, with the word 'poly' meaning 'many', and 'nomials' meaning 'terms'.

$$\underbrace{6x^2y}_{\text{term}} + \underbrace{14y^3}_{\text{term}} - \underbrace{7x}_{\text{term}} + \underbrace{15}_{\text{term}}$$

A polynomial:

The coefficient of the first term is 6.

The coefficient of the second term is 14.

A *term* is a single expression, which may contain multiplication or division. Monomials have one term (e.g. $4xy$), binomials have two terms (e.g. $3x - 5y^2$) and trinomials have three terms (e.g. $3x^2 - 5x + 2$).

The degree of a term is the sum of exponents of all of the variables. The degree of a polynomial is the highest degree of all of its terms.

A polynomial with degree zero is called a 'constant' (e.g. 3) while a polynomial with degree 1 is called 'linear' (e.g. $7x + 2$). A polynomial with degree 2 is called a 'quadratic' (e.g. $-5x^2 + 3x - 8$). A 'cubic' polynomial has a degree of 3, a 'quartic' polynomial has a degree of 4 and a 'quintic' polynomial has a degree of 5.

Like terms are terms with the exact same variable parts. Like terms are simplified by adding/subtracting their coefficients. For example, $3x^2 - 5x^2 = -2x^2$.

Expanding an expression means using the distributive property of multiplication.

For example, $-5(3x^2 - 5x + 1) = -15x^2 + 25x - 5$.

Solving an equation or inequality requires performing the same operation on both sides of the equality/inequality. In order to isolate the variable, you perform the 'opposite operation'.

The biggest difference between an equation and an inequality is that if you multiply or divide each side of an inequality by a negative number you need to flip the sign of the inequality.

You usually represent the solution of an inequality on a number line. If the inequality is 'greater than' or 'less than', you use an unshaded circle. If the inequality is 'greater than or equal to' or 'less than or equal to', you use a shaded (filled-in) circle.

Unit review

criterion A

Launch additional digital resources for this chapter

Key to Unit review question levels:

Level 1–2   Level 3–4   Level 5–6   Level 7–8

1   Classify each polynomial in as many ways as possible (number of terms, degree, constant/linear/quadratic/cubic/quartic/quintic).

   a $3t^2 + 5t - 2$

   b $4w - 9$

   c $m^2y^6 + my^5 - 3m^2y^2 + 8my^4$

   d $2x^3$

2   Simplify these expressions.

   a $-4w^3y^4 + 7w^2y + 9w^2y - w^3y^4$

   b $8m^3 - 7m + 2 + 10m - 11m^3 - 6$

   c $4gh^2 + 8g^2h - 6gh^2 - gh^2 - 9g^2h$

   d $1 - 3f - 8f - 12f^2 + 7$

3   **Solve** these equations.

   a $3z + 5 = 14$

   b $d - 3 = 2d + 1$

   c $6w = -12$

   d $2 - z = 14$

4   **Solve** these equations.

   a $3w + 5 = 2w + 7$

   b $-6t + 5t + 3 = 4 + t$

   c $9r + 3 - 6r - 2 = -2r + 8$

5   **Solve** these inequalities and represent each solution on a number line.

   a $v + 6 > -1$ b $3g \leq -6$ c $2 \leq x - 4$ d $5 > \dfrac{x}{2}$

6   **Use** algebra to **justify** why the following number trick always works.

   • *Take an integer and add to it the next consecutive integer.*

   • *Add 7 to this result.*

   • *Divide your new number by 2.*

   • *Subtract the original number.*

   • *The result is 4.*

7 **Solve** these equations.

a $\dfrac{2b}{5} + 10 = 8$

b $\dfrac{2a}{3} + 3 = 13 - a$

8 **Solve** the inequalities and represent each solution on a separate number line.

a $3t - 4 < 6t + 5$

b $-2g + 7 \le 11$

c $2a - 12 > \dfrac{a}{2}$

9 **Justify** the following number tricks using algebra.

a Take any three consecutive integers and add the highest and lowest numbers. **Show** that the sum is always double the middle number.

b Take any four consecutive integers and add the highest and lowest numbers. **Show** that the sum is always equal to the sum of the two middle numbers.

c Take any five consecutive integers and add the highest and lowest numbers. **Show** that the sum is always double the middle number and the same as the sum of the other two middle values.

10 **Find** two numbers that differ by 5, where the sum of the smaller one and four times the larger one is 100.

11 **Solve** these equations.

a $5(g - 3) + 2 = 3(g + 1)$

b $\dfrac{4(x - 4)}{3} - 3x = x$

12 **Use** algebra to **justify** why the following number tricks always work.

a If you take a two-digit number and subtract the number created when you reverse its digits, you always get nine times the difference in the digits.

b Show that the mean of a set containing an odd number of consecutive integers is always the middle integer in the set.

13 **Find** three consecutive numbers such that three times the lowest one plus five times the middle one is equal to 39 less than ten times the largest one.

14 The Atbash cipher was used around 500 BC and maps the alphabet in reverse.

Plain	A	B	C	D	E	F	G	H	I	J	K	L	M	N	O	P	Q	R	S	T	U	V	W	X	Y	Z
Cipher	Z	Y	X	W	V	U	T	S	R	Q	P	O	N	M	L	K	J	I	H	G	F	E	D	C	B	A

a **Find** an encryption function for the Atbash cipher.

b **Write down** a message and encode it with this cipher.

c **Find** the inverse of the encryption function. **Show** that the inverse function decodes your message.

15 You are creating a number trick of your own. You will ask a volunteer to select a number and then you will:

- ask them to *multiply the number by two.*
- choose an even number and ask your volunteer to *add this number to the one in your head.*
- say *divide the result by 2 and subtract your original number.*

What is the result? **Justify** your answer using algebra.

16 In pairs, try the following card trick to see if you and your partner can figure out how it works using algebra.

Using a standard deck of 52 playing cards, turn the top card of the deck face up.

Begin counting from the face value of this card, turning cards up from the deck until you have counted to 14.

- For the picture cards: you count jacks as 11, queens as 12 and kings as 13. Aces count as 1.
- For example, if you turned over a 6 first, you would turn over eight more cards to put in that pile ($14 - 6 = 8$).

Next, you form three more piles using the exact same process.

Then turn all four piles face down.

Discard leftover cards in a separate pile.

Have a peer pick up two of the piles and put them in the discard pile – without you seeing which piles they removed so you do not know the cards.

Then have your peer turn up the top card of one of the two piles remaining.

Deal 22 cards from the discard pile and place off to the side.

Remove the number of cards from the discard pile that corresponds to the face value of the card your peer turned up.

Count the number of cards left in the discard pile. It will equal the face value of the card on the top of the last pile.

Summative assessment

In this summative task, you will analyse a mathematical puzzle and then create one of your own.

In class portion: The Original Flash Mind Reader

Search on the internet for the 'Original Flash Mind Reader'. You will play the game several times and look for a pattern. You will then figure out how the game works and justify the results that you see.

Play the game

1 Your teacher will say if you are to play alone or with a partner. Pick any two-digit number. Add the digits together and subtract that result from your original number. Write this result in a table like the one below. Find this number on the game's grid and look at the symbol next to it. Write this in the table and then press the blue circle.

2 Repeat this procedure several times.

Original number	Sum of digits	Original number minus the sum of the digits	Symbol
Example: 34	7	27	

How does it work?

3 Write down the pattern that you see with your results.

4 Explain how you know which symbol will appear next.

5 Justify, using algebra, the rule you have discovered.

6 Verify your rule for 2 more examples.

Create your own puzzle/trick

You will now create two number puzzles and/or tricks that can be justified using algebra. Both of your puzzles/tricks must require the use of several steps.

1 The first one must be justified using one or more algebraic expressions. You must create the trick/puzzle and then justify why it works using algebra, showing all working.

2 The second one must involve an equation which can be solved to find the original value. You must create the trick/puzzle and then justify why it works showing all working.

3 Once the tricks/puzzles have been assessed by your teacher and any necessary modifications made, your teacher will instruct you to do one of the following:

- Write your puzzle/trick on card stock supplied by your teacher with the question on the front and solution on the back.

- Create a digital version of the puzzle on slides to be compiled by your teacher.

- Prepare a short instructional video explaining the mathematics behind your puzzle or trick, using an app like Explain Everything or Show Me. In this video you would state the puzzle and allow your peers to pause the video to attempt the puzzle before watching the rest of the video to see if they are correct.

You may even have a class where you try and play/figure out each other's puzzles and tricks.

Reflect and discuss 10

Can every puzzle be solved?

Can every trick be explained?

5 2D and 3D geometry

The structures created by humans and structures in nature provide a wealth of examples for your study of two- and three-dimensional geometry. These same shapes are also found in a variety of other contexts, from art to scientific innovation. Perimeter, area and volume are common mathematical concepts, but their applications to these new contexts could inspire you use them in even more varied ways.

Personal and cultural expression

Appreciation of the aesthetic

Art involves creation and inspiration, but that does not mean that art can't also involve calculation. Is "good art" an accident or can it be measured and calculated? How does knowledge of mathematics affect artistic expression and interpretation? The study of area and volume can help you to help analyse works of art but may also inspire you to create your own.

The 'Square head' in Nice was the first large-scale sculpture to be made into a usable building. It is 28 meters high, with seven floors. How would you go about planning art on such a large scale?

This memorial in Berlin was built in remembrance of Jewish victims of the Holocaust. It consists of 2711 differently-sized concrete slabs arranged on a slight slope.

How could you calculate the amount of concrete needed to create each block? What effect might the different sizes have on the viewer?

Scientific and technical innovation

Scientific and technological advances

Technology is often associated with improved communications and computing but it has also provided hope for many people.

The invention of the 3D printer has literally given a hand to people who have undergone amputation. A 3D printed prosthetic limb can cost considerably less than a traditional prosthesis, making it a much more affordable option.

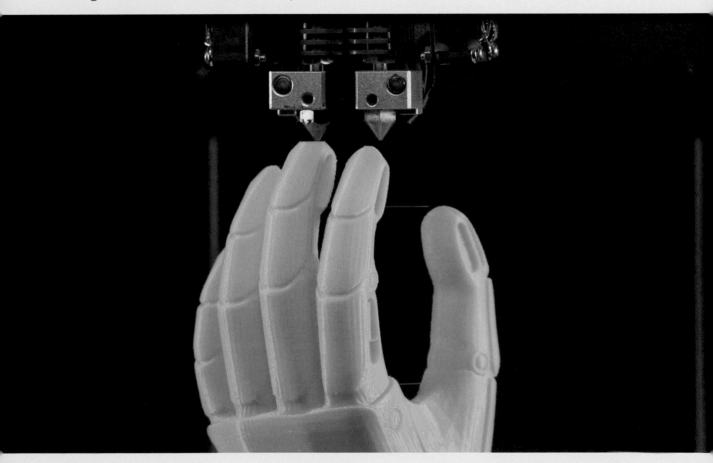

Bioprinting also has the power to change lives. Using patients' own skin cells, synthetic skin can be created for burn victims, eliminating the need for skin grafts, which often create scars and new wounds. This new synthetic skin has also been shown to speed up recovery time, offering hope to patients all over the world.

In both of these technologies, being able to calculate the surface area and volume of the required shapes can help to minimize the amount of material used and shorten production time. It can also help tailor the treatment to each individual, increasing the likelihood of success.

5

2D and 3D geometry
Human and natural landscapes

Related concepts: Generalization and Measurement

Global context:

In this unit, you will explore natural and human-made landscapes as part of your study of the global context **orientation in space and time**. As you analyse the formation of these structures, you will apply the concepts of perimeter, area, surface area and volume. Sometimes, it may even be hard to distinguish which landscapes are natural and which ones are man-made.

Statement of Inquiry:

Generalizing relationships between measurements can help explore the formation of human and natural landscapes.

Objectives

- Finding the perimeter and area of 2-dimensional shapes, including circles and trapezoids
- Finding the surface area and volume of 3-dimensional shapes, including prisms
- Solving problems involving 2D and 3D shapes

Inquiry questions

F What is a measurement?

C How are area and volume related?
How do measurements help define spaces?
How do we generalize relationships between measurements?

D Which exhibit more order, natural or human landscapes?
Do humans mimic nature or does nature mimic humans?

ATL1 Transfer skills

Apply skills and knowledge in unfamiliar situations

ATL2 Communication skills

Make effective summary notes for studying

You should already know how to:

- round numbers
- find the perimeter and area of simple 2D shapes
- convert between units

- convert fractions to decimals
- multiply mixed numbers/fractions
- solve problems involving percentages
- solve equations

Introducing 2D and 3D geometry

You are surrounded by landscapes that are both natural and human-made. From mountains to cities to caves to worlds beyond our planet, there is a relationship between that which is original and that which is not. Sometimes, the lines between the two are blurred, so much so that you cannot necessarily tell which is the product of centuries of natural processes and which was built over the course of a few years.

For example, in the picture to the left, are the geometric formations natural or artificial? How was it formed?

How and why natural and human landscapes were created is the focus of this unit.

Reflect and discuss 1

- Do you think the structures in the above picture are a natural phenomenon or an artificial one? Describe what you think it is and explain your reasoning.

- Are natural formations random, or do they demonstrate some kind of order? Explain using three specific examples.

- Are human structures designed to mimic nature, enhance it, neither or both? Explain using a specific example.

In this unit, you will explore the formation of both natural and human landscapes and discover some of the reasons why they have the shapes they do. You may be surprised to find out which structures follow scientific principles and which ones are 'simply made that way'.

2D figures

Trapezoid

A *trapezoid* is a quadrilateral with just one set of parallel sides. Last year, you looked at the trapezoid as a compound shape, made up of a rectangle and two triangles. This year, you will explore the area of a trapezoid as a single shape.

Investigation 1 – Area of a trapezoid

ATL1

The trapezoid below is called an *isosceles trapezoid* because the two sides that are not parallel are equal in length. Sides a and b are parallel and the height is labelled h.

criterion **B**

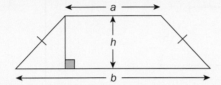

1 Create an isosceles trapezoid (either by hand on paper or using dynamic geometry software). As long as the two bases are parallel, the other two sides can be any length that you want.

2 Make an exact copy of it. If using paper, cut out the original trapezoid and trace it, then cut out the copy.

3 Rotate the copy 180°.

4 Translate (move) the copy so that it connects to the original completely on one side. What shape do you have now?

5 Knowing the formula to find the area of this new shape, write it in terms of a, b and h.

6 Since the shape is composed of two trapezoids, write down the formula for the area of one trapezoid.

7 Can you think of two other ways you could write this formula? Write them down.

8 Research other types of trapezoids and draw one that is not isosceles. Complete the same steps and see if the process and formula hold true for other types of trapezoids. Alternatively, your teacher may tell students which specific trapezoid they are to create at the beginning of the investigation and then you will compare your answers with the answers of your classmates who used a different type of trapezoid.

> Remember BEDMAS, so make sure you include a set of brackets.

Reflect and discuss 2

- Some people say, 'Finding the area of a trapezoid is like finding the area of a rectangle that has a side length equal to the average of the side lengths of the trapezoid'. How is this represented in your formula?

- Show how you can obtain the same formula for the area of a trapezoid by dividing it into smaller pieces and adding their areas together.

- Name three real-life objects that are shaped like trapezoids.

Example 1

(Q) Find the area of this trapezoid.

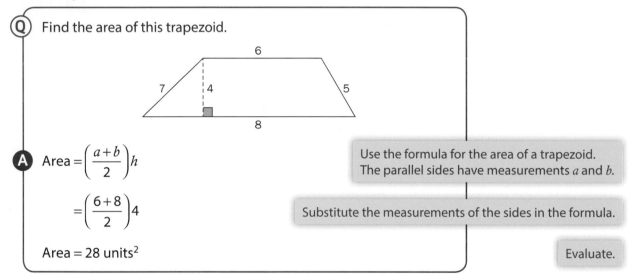

(A) Area $= \left(\dfrac{a+b}{2}\right)h$

> Use the formula for the area of a trapezoid. The parallel sides have measurements a and b.

$= \left(\dfrac{6+8}{2}\right)4$

> Substitute the measurements of the sides in the formula.

Area $= 28$ units^2

> Evaluate.

Regular polygons

A *regular polygon* is one whose sides all have the same length and whose interior angles are all congruent (equal in measure). The perimeter of a regular polygon, then, is simply the length of any side multiplied by the number of sides. In a regular polygon, the perpendicular distance from the center to any side is called the *apothem*.

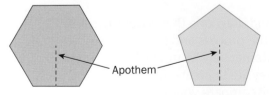

Apothem

The area of a 2-dimensional figure can be defined as the amount of space it covers. You already know how to find the area of regular polygons like squares and equilateral triangles, but what happens when the figure has more than four sides?

Investigation 2 – Area of regular polygons

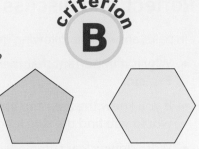

criterion **B**

1 Given a regular polygon (like the ones on the right), how could you divide it into shapes whose area you know how to calculate? Show your answer using a diagram.

2 Discuss your answers with a peer or two. What method would work to divide *any* regular polygon into smaller shapes?

3 Complete a table like the following, showing all of your work.

Polygon	Name	Number of sides	Number of isosceles triangles	Area calculation of polygon
s = 16 cm a = 11 cm				
s = 29 mm a = 25 mm				
s = 10 cm a = 12 cm				
s = 15 m a = 23 m				

4 Write down a formula to find the area of a regular *n-sided* polygon with sides of length *s*, and an apothem that measures *a*.

Reflect and discuss 3

In pairs, answer the following questions:

- Why is it important that the polygons in Investigation 1 be regular? Explain.

- If you knew the area and apothem of a regular pentagon, how would you find the side length? Show your work.

ATL2
- Write down clear summary notes regarding how to find the perimeter and area of trapezoids and regular polygons. Be sure to include formulas and to define what each variable means. When done, compare your notes with those of a peer.

Did you know?

While regular polygons can often be found in human structures, they are also seen in natural landscapes as well. The morning glory flower is actually shaped like a regular pentagon. Because they grow quickly, even in poor soil, morning glory plants are sometimes used to cover walls of human landscapes, such as buildings, to reduce both heating and cooling costs. A single morning glory plant can produce over 60 000 flowers at a rate of 300 flowers per day!

Example 2

Q The side of a typical morning glory flower measures 3 cm and its apothem measures 2.1 cm.

Find the area that one morning glory plant could cover in flowers in one day, assuming the plant produces 300 flowers that day.

A

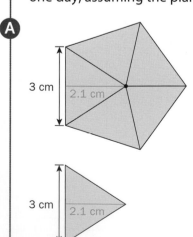

Divide the shape into five small triangles.

Find the area of one of the triangles.

▶ Continued on next page

Area of one triangle $= \dfrac{1}{2}$ (base × height)

$$= \dfrac{1}{2}(3 \text{ cm} \times 2.1 \text{ cm})$$

$$= 3.15 \text{ cm}^2$$

Area of pentagon $= 5 \times 3.15 \text{ cm}^2$
$$= 15.75 \text{ cm}^2$$

Since there are five triangles, find the area of the pentagon by multiplying the area of the triangle by 5.

If you remember the formula developed for the area of a pentagon, $A = n \times \dfrac{bh}{2}$, then you can just substitute the values of n, b and h into it.

Area after one day $= 15.75 \text{ cm}^2 \times 300$
$$= 4725 \text{ cm}^2$$

Since 300 flowers can be produced in one day, multiply the area of one flower by 300.

Practice 1

1 Find the perimeter and area of the following polygons. Assume all polygons apart from the trapezoids are regular polygons. Round your answers to the nearest tenth.

a
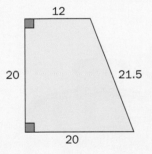
12
20
21.5
20

b
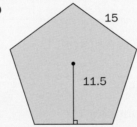
15
11.5

c
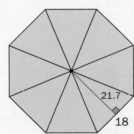
21.7
18

d
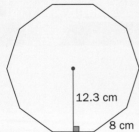
12.3 cm
8 cm

e
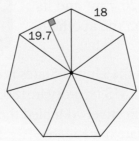
19.7
18

f
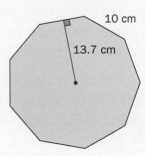
10 cm
13.7 cm

g
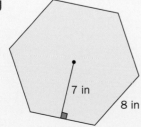
7 in
8 in

h
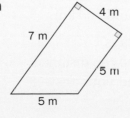
4 m
7 m
5 m
5 m

▶ Continued on next page

2 Show that the area of a square calculated using the apothem is the same as the area using the standard formula.

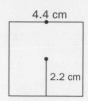

4.4 cm

2.2 cm

3 Make a copy of this table. Assuming each figure is a regular polygon, find the missing value.

Round answers to the nearest tenth where appropriate.

Polygon	Side length (cm)	Apothem (cm)	Area (cm²)
pentagon	6		61.9
hexagon	10		259.8
octagon		5	82.8
decagon	20		3077.7
hexagon		12	498.8
pentagon		30	3269.4

4 The Rich-Twinn Octagon House is a historic house in Akron, New York, and is on the National Register of Historic Places.

a Sketch a diagram of the floor plan for the ground floor.

b The area of the floor plan is 120.7 m² and its perimeter measures 40 m. Find the apothem of the octagonal floor plan.

ATL1 **5** The natural landscape that forms the United States is divided into 50 states. North Dakota became a US territory in 1803 as part of the Louisiana Purchase, which effectively doubled the size of the United States at that time. The state has roughly the shape of a trapezoid with its parallel sides measuring approximately 580 km and 500 km.

a If the total area of North Dakota is approximately 183 000 km², find the distance between its parallel sides. Round your answer to the nearest ten.

b Is this a natural or a human landscape? Explain.

▶ Continued on next page

6 Salar de Uyuni in Bolivia is the largest salt flat on the planet. Over 30 000 years ago, the region was part of a prehistoric lake. Climactic change, including an increase in temperatures, caused the lake to dry up, leaving behind tiles of salt that are roughly hexagonal in shape due to their crystalline structure.

The evaporating groundwater beneath it produces a continuous supply of lithium-rich salt, even though the area can sometimes be temporarily covered with water after a rainfall. Over half of the lithium in the world's lithium-ion batteries comes from this natural landscape.

a If the side of each hexagonal tile measures 30 cm and its apothem measures 26 cm, find the area of one hexagonal salt tile.

b The salt flats have an approximate area of 10 600 km². Find the number of hexagonal tiles that could cover such an area.

7 The Pentagon in Washington, DC, is one of the world's largest office buildings. It took approximately a year and a half to build, but was designed to use a minimal amount of steel. The structure is actually two pentagons: the outer pentagon 'ring' houses the office space and the inner pentagon is a courtyard containing a garden.

a The inner pentagon has sides that measure 109.7 meters and an apothem measuring 75.5 meters. Find the area of the inner courtyard.

b If the outer pentagon has sides that measure 280.7 meters and an apothem that measures 193.2 meters, find the area that is dedicated as office space.

Circles

How do you find the perimeter of an object that has no sides and does not contain any line segments? The perimeter of a circle is called the *circumference,* and finding the circumference of a circle is not as difficult as it may sound. The next investigation will help you generalize the relationship between the circumference of a circle and one of its measurements.

Investigation 3 – Circumference of a circle

criterion B

1 Take five cylindrical objects and trace around the base of each one to produce five different circles. Measure the diameter of each circle and fill in the appropriate column in a table like the one below.

Circle	Diameter (d)	Circumference (C)	$\frac{C}{d}$
1			
2			
3			
4			
5			

2 Take a piece of string and place it around the circumference of a circle as closely as you can. Measure the length of the string that represents the circumference.

3 Repeat this procedure for the rest of the circles and fill in the appropriate column in your table.

4 Divide the circumference by the diameter $\left(\dfrac{C}{d}\right)$ for each circle. Write your results in the last column.

5 Compare your answers to those of at least two classmates. What do you notice about all of your answers in the last column? Explain.

6 Research what this number is called and write down its value, rounding to the eighth decimal place.

7 Generalize the relationship between the circumference of a circle and its diameter, hence find the formula to calculate the circumference of a circle.

8 Draw another circle and verify your formula for another case.

Reflect and discuss 4

- If you knew the radius of a circle, how could you calculate its circumference? Explain.

- Write down a formula for the circumference of a circle, given that you know its radius.

As with all 2D geometric shapes, you can find the area of a circle. However, generalizing the relationship for the area of a circle is quite different than for a shape whose sides are straight line segments. This is the focus of the next investigation.

 ## Investigation 4 – Area of a circle

criterion **B**

1 Draw a circle on paper and cut it out.

2 Then cut the circle into quarters and arrange them as shown here:

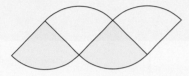

3 Copy the arrangement in your notebook.

4 Take the quarters and cut each of them in half. Arrange the pieces similarly to how you did in step **2**. Copy the arrangement in your notebook.

5 Repeat the procedure one more time, cutting the pieces in half, arranging them and copying the arrangement in your notebook.

6 What shape seems to be formed by the sixteen pieces? How do you find the area of this shape?

7 Determine the measure of the height of the shape formed by the sixteen pieces.

8 Determine the measure of the base of this shape in terms of the radius r.

9 Write down a formula to calculate the area of a circle, given that you know its radius.

Reflect and discuss 5

- Explain why the area of any figure is always measured in *square* units (e.g. cm^2, m^2).

ATL2 - Write down clear summary notes on how to find the circumference and area of a circle. Be sure to include formulas and to define what each variable means. When done, compare your notes with those of a peer.

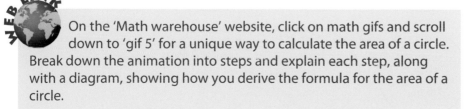

On the 'Math warehouse' website, click on math gifs and scroll down to 'gif 5' for a unique way to calculate the area of a circle. Break down the animation into steps and explain each step, along with a diagram, showing how you derive the formula for the area of a circle.

Example 3

Q Nearly 25 meters below sea level, off the coast of Japan, are underwater circles like the one on the right.

Created by male puffer fish to attract mates, these circles can measure up to 1.8 meters in diameter.

a Find the circumference of one of these circles if its radius measures 0.9 m.

b Find the area of one of these circles with a radius of 0.9 m.

Leave your answers in terms of π.

$C = 2\pi r$

$\quad = 2\ \pi(0.9)$

$\quad = 1.8\pi$ meters

The circumference is 1.8π meters.

> The circumference of a circle can be calculated using the formula $C = \pi r$.

$A = \pi r^2$

$\quad = \pi(0.9)^2$

$\quad = 0.81\pi\ m^2$

The area is 0.81π square meters.

> The area of a circle can be calculated using the formula $A = \pi r^2$.

> To see the puffer fish create one of these circles, search online for 'BBC Four puffer fish crop circles'.

Practice 2

1 Find the circumference and area of the following circles. Use 3.14 for the value of ≠, and round your answers to the nearest hundredth.

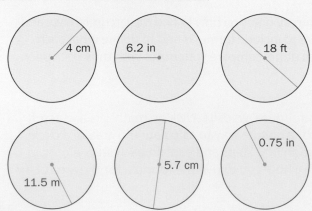

4 cm

6.2 in

18 ft

11.5 m

5.7 cm

0.75 in

▶ Continued on next page

2 One of the largest circular labyrinths in the world can be found at Reignac-sur-Indre in France. With a diameter of approximately 100 m, every year the maze grows back a little different than the year before.

a If you could start at some point on the outside and walk straight through the middle of the labyrinth to arrive on the other side, how much shorter would it be than to walk halfway around the circumference and arrive at the same point? Show all of your working.

b How would you describe this landscape?

ATL1 3 In order to host track and field competitions, a 400-meter track needs to be created. A typical Olympic track looks like the following:

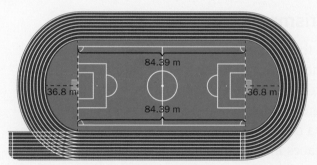

a Show that the runner who runs in the inside lane runs 400 m.

b All other runners in the 400 m race start a little further ahead since their lane is a little longer. If the radius of lane 2 (the one beside the inside lane) is 37.92 m, find how much of a stagger (head start) the lane 2 runner should be given. Round your answer to the nearest hundredth.

3D figures

Some of the most spectacular 3D shapes can be found in crystals that form naturally all over the world. The Giant Crystal Cave in Mexico houses some of the planet's largest selenite crystals. The cave is located above an underground magma chamber, which heated groundwater full of minerals that formed the crystals as the water evaporated. Because the cave was sealed off from the elements, the crystals were able to grow for over 500 000 years, much longer than most other crystals. The result is that some of these crystals are shaped like hexagonal prisms.

Surface area of regular prisms

Every 3D shape can be flattened to 2 dimensions. As long as all of the faces are connected somehow, we call these 2-dimensional versions *nets*. The *surface area* of a 3-dimensional shape is the area of all of its faces combined together. Creating a *net* is an efficient way to determine the surface area of a 3D shape, since you can flatten the shape to see all of its faces and then add the areas of the individual sections.

Investigation 5 – Surface area of regular prisms

criterion

B

Pairs

In pairs, determine the nets of each of the shapes in the table and hence write the formula for the surface area of each shape. To determine the surface area:

- draw the net of each 3D shape
- label the dimensions of each section
- find the area of each section
- total the areas of all of the sections
- write down the formula for the surface area of the prism
- simplify this formula, if necessary.

▶ Continued on next page

Your teacher will provide you with manipulatives of 3D prisms that can be broken down into nets. Alternatively, you can look at virtual manipulatives.

WEB LINK

Search online for 'Annenberg Learner interactives prisms' where you can see how the net unfolds as a visual to help you determine the formula.

Create a table with the following headings, making sure to have enough room to draw the net and label all dimensions clearly.

3D Solid	Dimensions & shapes of sections needed to determine surface area (SA)	SA Formula
Triangular prism		
Rectangular prism		
Pentagonal prism		
Hexagonal prism		
Octagonal prism		
Decagonal prism		
n-sided polygonal prism		

Reflect and discuss 6

- How is finding the surface area of a 3-dimensional shape similar to finding the area of a compound 2-dimensional shape? Explain.

- What do you think is more important: knowing the surface area formulas, or being able to reproduce them through drawing a net? Explain.

ATL2

- Write down clear summary notes on how to find the surface area of a regular prism. Be sure to include formulas and to define what each variable means. When done, compare your notes with those of a peer.

Example 4

Q Find the surface area of the hexagonal prism crystal described in the Giant Crystal Cave. The hexagon has side lengths of 1 m, apothem length of 0.87 m and the prism has a length of 11 m. Round your answer to the nearest tenth.

A

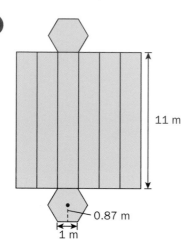

Draw the net and indicate its dimensions.

The area of one side is: $1 \text{ m} \times 11 \text{ m} = 11 \text{ m}^2$

The total area of the sides is: $11 \text{ m}^2 \times 6 = 66 \text{ m}^2$

Multiply the area of one side by the number of sides.

The area of one of the hexagonal bases is:

$6 \times \left(\dfrac{1 \times 0.87}{2} \right) = 2.61 \text{ m}^2$

Area of the triangle is $\frac{1}{2}(bh)$, and there are six of them in the hexagon.

The area of the two hexagons is: $2 \times 2.61 \text{ m}^2 = 5.22 \text{ m}^2$

The total surface area of the crystal is

There are two hexagons in the hexagonal prism.

$66 \text{ m}^2 + 5.22 \text{ m}^2 = 71.22 \text{ m}^2 \approx 71.2 \text{ m}^2$

Add the areas together to find the total surface area.

Practice 3

1 Find the surface area of each of these prisms.

a
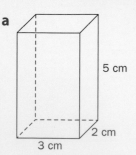
5 cm
2 cm
3 cm

b
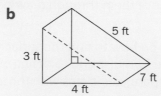
5 ft
3 ft
4 ft
7 ft

c
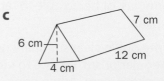
7 cm
6 cm
4 cm
12 cm

▶ Continued on next page

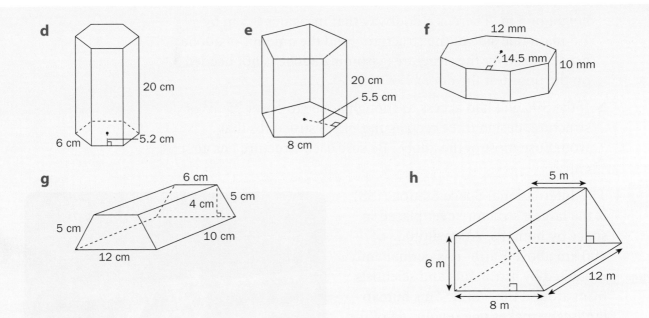

d 20 cm, 6 cm, 5.2 cm

e 20 cm, 5.5 cm, 8 cm

f 12 mm, 14.5 mm, 10 mm

g 6 cm, 4 cm, 5 cm, 5 cm, 10 cm, 12 cm

h 5 m, 6 m, 8 m, 12 m

2 Assuming each figure is a regular prism, find the missing values in the following table. Round answers to the nearest tenth where appropriate.

Prism	Side length (cm)	Apothem (cm)	Height of prism (cm)	Surface area (cm²)
Pentagonal	10	6.9	18	
Hexagonal	7	6.1	7	
Heptagonal	12	12.5	21	
Octagonal	20	24.1		4656
Decagonal	16	24.6		6176

3 Montezuma Castle is a dwelling constructed in a natural alcove about 25 m off the ground on the side of a sheer limestone cliff. It was made by the Southern Sinagua (Spanish for 'without water') people of Arizona, USA, who occupied it from approximately 1100 to 1450 AD. It is a series of rooms spread over five levels that served as storage and apartments for approximately 50 people. The structures, mostly rectangular prisms, were made of stone and wood and covered in mud (called 'adobe'). The alcove provided natural protection from the elements, rivals and flooding.

a The 'tower' in the middle of the structure is a rectangular prism with side lengths of 1 m and 2 m and a

▶ Continued on next page

height of 3 m. There is a doorway that measures 0.5 m by 1.5 m on one side of the structure. Find the amount of adobe needed to cover the structure (assuming adobe is not needed on the floor nor the door).

b If the Sinagua had access to enough adobe to cover 75 m² of structure, design three rectangular prism structures that would use most of the adobe. Be sure each structure has an entry.

ATL1 4 The International Space Station (ISS) is the largest structure ever placed in space by humans. At an altitude of 400 km above Earth, it is permanently inhabited by astronauts and scientists from around the world. This human landscape makes one revolution of the planet every 90 minutes and can be seen in the sky with the naked eye. The pieces for the station were taken into space where they were assembled, and the station is now as big as a football field, complete with two bathrooms and a gym.

One of the components (P3) is a hexagonal prism with a height of 7 m. Each hexagonal base has a side length of 2.7 m and an apothem of 2.4 m.

a If P3 needs to be completely covered in solar panels, find the area of panels that must be brought into space.

b As more pieces are added to the ISS, the panels on the hexagonal bases of P3 will be removed since the hexagonal bases will be attached to structures on each end. What percentage of the total area do these bases represent?

c If it had been decided to use a pentagonal prism instead of a hexagonal one, find how long the prism would have to be in order to maintain the same area of solar panels as the hexagonal version. Assume the side lengths and apothem are the same as before.

To see when the ISS will be visible from where you are, search online for 'ISS sightings'.

Volume

Shipping container architecture (or 'cargotecture' or 'arkitainer') is a recent method of building everything from homes to offices to restaurants, like the ones pictured here from Contenedores Food Place in Colombia.

Shipping containers have been slowly introduced into human landscapes because they are sturdy and inexpensive. It is often too expensive to return them to their origin after delivery, so repurposing them is now considered good environmental practice. How much space do they actually contain, and does that represent truly liveable space? Will the cargotecture movement eventually reduce our own need for space or will it be just one of many alternatives?

Volume of rectangular prisms

The volume of a rectangular prism can be calculated using $V = l \times w \times h$. As with area, it may be necessary to divide a 3D shape into smaller shapes in order to calculate its volume.

Example 5

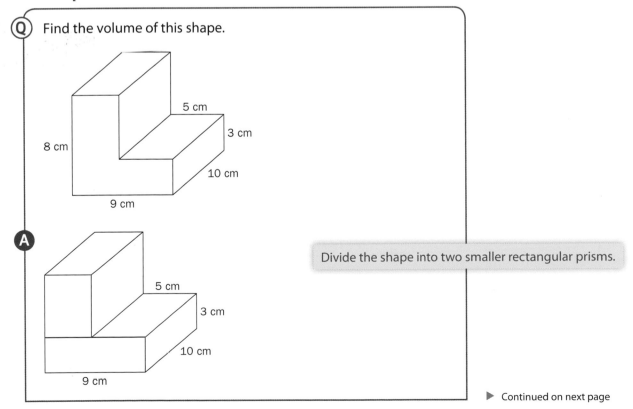

(Q) Find the volume of this shape.

A

Divide the shape into two smaller rectangular prisms.

▶ Continued on next page

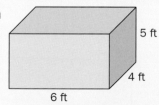

5 cm

5 cm

4 cm

3 cm

10 cm

9 cm

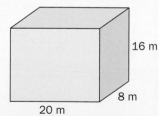

Calculate the missing values.

Since the maximum height of the shape is 8 cm and the height of the lower section is 3 cm, the height of the upper portion must be 5 cm.

Since the length of the whole shape is 9 cm and the length of a portion of the lower section is 5 cm, the rest of that length is 4 cm.

Upper prism:

$V = l \times w \times h$
$\quad = 4 \text{ cm} \times 10 \text{ cm} \times 5 \text{ cm}$
$\quad = 200 \text{ cm}^3$

Calculate the volume of each section individually. The volume of a rectangular prism is found by using $V = l \times w \times h$.

Lower prism:

$V = 9 \text{ cm} \times 10 \text{ cm} \times 3 \text{ cm}$
$\quad = 270 \text{ cm}^3$

The total volume is 470 cm³.

Add the volume of the upper prism to that of the lower prism.

Practice 4

1 Find the volume of these rectangular prisms. Round answers to the nearest tenth where necessary.

a

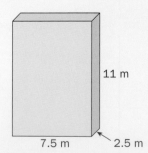

5 ft

4 ft

6 ft

b

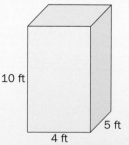

10 ft

5 ft

4 ft

c

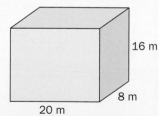

16 m

20 m

8 m

d

11 m

7.5 m   2.5 m

e

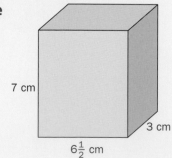

7 cm

3 cm

$6\frac{1}{2}$ cm

f

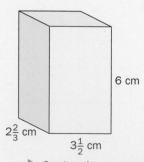

6 cm

$2\frac{2}{3}$ cm

$3\frac{1}{2}$ cm

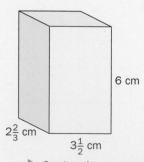

▶ Continued on next page

2 Assuming each figure is a rectangular prism, find the missing values in the following table. Round answers to the nearest tenth where necessary.

Length (cm)	Width (cm)	Height (cm)	Volume (cm³)
12	4		240
9		7	1260
	4.9	6.5	78
4.1	7.6		23.4
18.1	12.5	6	
	4.8	7.2	25.6

3 Find the volume of the following shapes. Round answers to the nearest tenth where necessary.

a

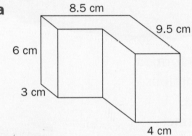

8.5 cm
9.5 cm
6 cm
3 cm
4 cm

b

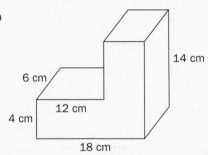

14 cm
6 cm
12 cm
4 cm
18 cm

c

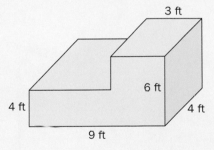

3 ft
6 ft
4 ft
4 ft
9 ft

d

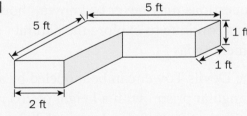

5 ft
5 ft
1 ft
1 ft
2 ft

e

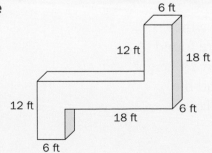

6 ft
12 ft
18 ft
6 ft
12 ft
18 ft
6 ft

f

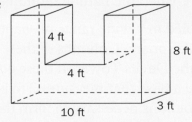

4 ft
8 ft
4 ft
10 ft
3 ft

▶ Continued on next page

4 Pyrite, often called fool's gold because of the number of gold miners that mistook it for real gold, can solidify in a wide range of configurations. Formed hundreds of millions of years ago when sulfur reacted with iron in sedimentary rock, it can be found all over the world. However, few pyrite deposits are as spectacular as those in Navajun, Spain. When the temperature and pressure are just right, then pyrite solidifies into perfect cubes.

a What volume of pyrite is contained in a cube that measures 8 cm on each side?

ATL1 **b** What are the dimensions of a cube of pyrite with a volume of 1331 cm^3?

c A pyrite cube has a base area of 5.76 cm^2. If the price of gold is 785 euros per cm^3, find how much a miner would think this cube of 'gold' would be worth.

5 Scientists agree that the Devils Tower in Wyoming was formed when magma forced its way in between other rock formations. What they cannot agree on is how that process occurred, and how far the lava traveled before cooling. Is it an extinct volcano? Was it formed underground and then exposed by erosion? The answers to these questions may never be known, though scientists continue to explore the natural monument today.

a If the Devils Tower can be modeled by a rectangular prism, with a height of 265 m and a base shape with a length of 300 m and a width of 500 m, find the volume of magma that created the natural monument.

b The lava cooled in a wide range of shapes, one of which is a rectangular prism. Columns of these prisms are found throughout the Devils Tower. If these columns have a height of 100 m and a base that measures 2 m by 2 m, find the number of columns necessary to form the Devils Tower.

▶ Continued on next page

Volume of other prisms

The volume of any prism can be calculated by taking the area of the *base shape* and multiplying by the height. The base shape is the shape that has been repeated in order to produce the prism. Note that it is not necessarily the side that the prism sits on.

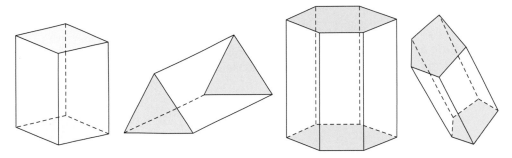

Investigation 6 – Volume of solids

criterion B

1 Create a copy of this table and fill it in.

Name of 3D shape	Name of base shape	Area formula of base shape	Formula of 3D shape
Triangular prism			
Rectangular prism			
Pentagonal prism			
Hexagonal prism			
Octagonal prism			
n-sided prism			
Trapezoidal prism			

2 Make a generalization about the relationship between area and volume of prisms.

Reflect and discuss 7
Reflect and discuss 7

- Explain how the formula for the volume of a rectangular prism ($V = l \times w \times h$) also represents 'the area of the base multiplied by the height'.

- Is it necessary to memorize the formulas from investigations 5 and 6? Explain your reasoning.

ATL2 • Write down clear summary notes on how to find the volume of a prism. Be sure to include a formula, and to define what each variable means. When done, compare your notes with those of a peer.

Example 6

Q Find the volume of this shape. Round your answer to the nearest whole number.

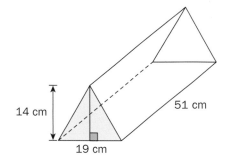

A The volume of the triangular prism is given by:

$V = $ area of triangle $\times$ height

$V = \dfrac{1}{2}bh \times$ height

> The base shape is a triangle, whose area is $\dfrac{1}{2}bh$. The volume then is that area multiplied by the height of the prism (51 cm).

$\quad = \dfrac{1}{2}(19\,\text{cm})(14\,\text{cm})(51\,\text{cm})$

$\quad = 6783\ \text{cm}^3$

The volume is 6783 cm³.

Practice 5

1 For each of the following shapes:

  i indicate the base shape and find its area

  ii find the volume of the 3D shape.

Show all of your working and round answers to the nearest tenth if necessary.

a

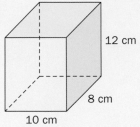

12 cm

8 cm

10 cm

b

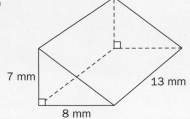

7 mm

8 mm

13 mm

c

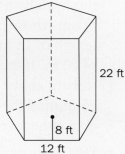

22 ft

8 ft

12 ft

d

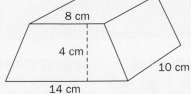

8 cm

4 cm

10 cm

14 cm

e

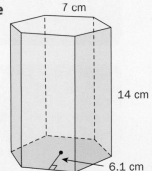

7 cm

14 cm

6.1 cm

2 Find the missing values in this table. Round answers to the nearest tenth where necessary.

Shape	Base dimensions	Height of 3D figure	Volume
Rectangular prism	length = ? width = 25 mm	35 mm	280 mm³
Triangular prism	base = 12 mm height = 7 mm	?	286 mm³
Pentagonal prism	side length = 8.2 m apothem = ?	10.1 m	451.8 cm³
Rectangular prism	length = 21 m width = ?	14.2 m	1300 m³
Hexagonal prism	side length = 5 mm apothem = 3 mm	?	112 mm³

▶ Continued on next page

3 Iceberg Alley, in Labrador, Canada, is known for the icebergs that float down from the Arctic every year. With an average of 250 icebergs per year, there have been some years with no icebergs, while others saw more than 2000 of them! The icebergs originate from the ancient glaciers of Greenland, believed to be up to 15 000 years old.

a One such iceberg can be modeled by a triangular prism whose base has a length of 15 m and a height of 20 m. If the height of the iceberg above the water is 25 m, find the volume of ice above the level of the water.

b Due to its density, 90% of an iceberg's volume is below the water. Find the volume of ice below the water.

4 Honeycombs are known to be one of the most efficient shapes in nature. Each of the angles of the honeycomb measures exactly 120 degrees. It has been determined that hexagons require the least amount of wax to make the hive, less than squares or triangles!

A typical honeycomb cell has sides that measure 2.7 mm and an apothem that measures 2.3 mm.

a If the depth of a typical cell is 11.3 mm, find the volume of honey that can fit in a single cell.

b An amateur beekeeper wants to make five bottles of honey. He will bottle his honey in containers in the shape of rectangular prisms with a height of 20 cm and a base with a length of 10 cm and a width of 8 cm. Find the number of honeycomb cells he will need to completely fill the five bottles.

Formative assessment

Cooling lava can take on a variety of shapes, the most common of which is the hexagonal prism. The Giant's Causeway is a collection of approximately 40 000 of these interlocking hexagonal basalt columns along the coast of Northern Ireland.

criterion D

1 Assuming a regular hexagon as the base shape of these columns, calculate the volume of basalt stone in a column that has an apothem of 19 cm, a side length of 22 cm and a height of 5 m.

2 Assuming this is an average size for the face shape, calculate the approximate area these structures cover.

3 Basalt can cool to form regular prisms with anywhere from 4 to 8 sides. Suppose one of the hexagonal prisms in question a) cooled to form a pentagonal prism instead. Find the length of its apothem if the perimeter and height of the column remains the same as in question a).

Research the answers to the following questions.

• Why do these columns mainly take the shape of hexagonal prisms?

• Where else on Earth does this phenomenon occur in natural landscapes?

• Are there any human landscapes that look like this? Why do you think that is?

• Do humans mimic nature or does nature mimic humans?

Unit summary

The area of a trapezoid can be found using the formula:

$$\text{Area} = \left(\frac{a+b}{2}\right)h,$$

where a and b are the measures of the parallel sides and h is the perpendicular distance between them.

A *regular polygon* is one whose sides all have the same measure and whose interior angles are all congruent (equal in measure). In a regular polygon, the perpendicular distance from the middle of the figure to any side is called the *apothem*.

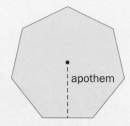
apothem

The area of a regular polygon with n sides of length s and an apothem measuring a can be calculated using

$$A = n\left(\frac{s \times a}{2}\right).$$

The circumference of a circle can be calculated using the formula

$$C = \pi d \text{ or } C = 2\pi r.$$

The area of a circle can be calculated using the formula

$$A = \pi r^2.$$

The *surface area* of a 3-dimensional shape is the area of all of its faces combined. Creating a *net* is an efficient way to determine the surface area of a 3D shape.

The surface area of a prism of height h whose base shape is a regular polygon with n sides measuring s and an apothem measuring a can be calculated using:

$$SA = n\left(\frac{s \times a}{2}\right)h$$

The volume of any prism can be calculated by taking the area of the *base shape* and multiplying by the height. The base shape is the shape that has been repeated in order to produce the prism.

The volume of a prism can be calculated using:

$$V = \text{area of base} \times \text{height}.$$

Unit review

criterion **A**

📑 **Launch additional digital resources for this chapter**

Key to Unit review question levels:

| Level 1–2 | Level 3–4 | Level 5–6 | Level 7–8 |

1 Find the perimeter and area of the following shapes. Round answers to the nearest tenth where necessary.

a

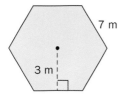

7 m

3 m

b

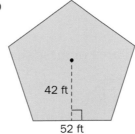

42 ft

52 ft

c

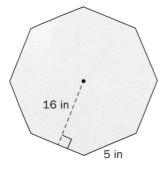

16 in

5 in

d

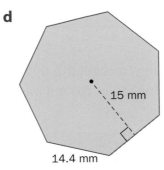

15 mm

14.4 mm

e

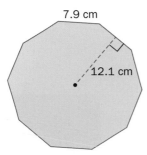

7.9 cm

12.1 cm

f

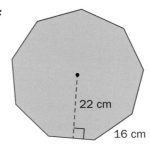

22 cm

16 cm

g

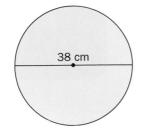

38 cm

h

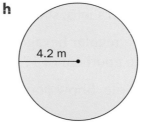

4.2 m

2 Find the volume and surface area of the following shapes.

a
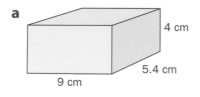
4 cm
5.4 cm
9 cm

b
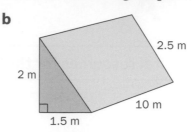
2.5 m
2 m
10 m
1.5 m

c
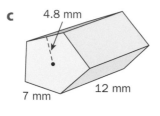
4.8 mm
7 mm
12 mm

d
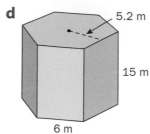
5.2 m
15 m
6 m

3 Find the perimeter and area of these shapes.

a
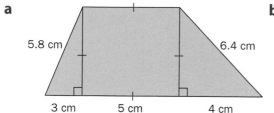
5.8 cm
6.4 cm
3 cm
5 cm
4 cm

b
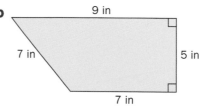
9 in
7 in
5 in
7 in

4 Assuming each figure is a regular polygon, find the missing values in the following table. Round answers to the nearest tenth where necessary.

Polygon	Side length (cm)	Apothem (cm)	Area (cm²)
Pentagon	9		92.3
Hexagon		10	144
Octagon	24		672.5
Decagon		8	1200

5 Antonio Gaudí was one of the most celebrated architects in Spain, designing houses, churches and even sidewalks. In many parts of Barcelona, you will find sidewalks that he designed in 1904 to represent what it would be like to walk on the bottom of the ocean, complete with shells, starfish and seaweed.

Each of Gaudí's tiles is a regular hexagon with sides measuring 20 cm and an apothem of 17.3 cm.

Find the area of a single tile. Show your working.

6 Find the surface area and volume of each shape.

a

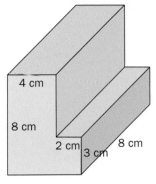

4 cm

8 cm

2 cm 3 cm 8 cm

b

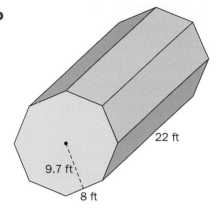

22 ft

9.7 ft

8 ft

c

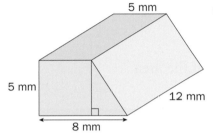

5 mm

5 mm

8 mm 12 mm

d

4 m

5 m 3 m

7 m 13 m

7 In a housing development, land is often broken into sections or parcels. These parcels can be kept together or sold individually for other people to build homes or otherwise develop. An example of parcels of land is given on the right.

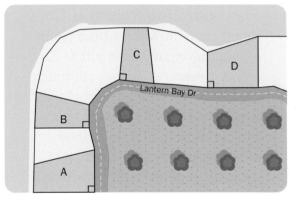

a Parcel A has two parallel sides that measure 20 m and 35 m. The perpendicular distance between these sides is 50 m. Find the area of the parcel.

b Parcel B has the same area as parcel A but its parallel sides measure 23 m and 27 m. Find the perpendicular distance between these parallel sides.

c Parcel C has parallel sides that measure 20 m and 29 m. The perpendicular distance between them is 47 m. Parcel C is being sold for 75 000 euros. Parcel D has the same perpendicular distance and price in total and per square meter, but one of its parallel sides measures 31 m. Find the measure of the other parallel side of Parcel D.

8 Crop circles are phenomena that began appearing in fields in the 1970s. At first attributed to aliens by some people, many have come forward to demonstrate how they are, in fact, a human landscape. The crop circle to the right was created overnight in a field in Wiltshire, England.

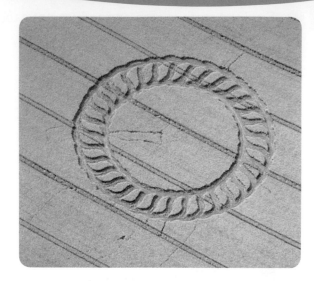

a If the diameter of the outermost circle measures 105 meters, find the area occupied by the whole crop circle.

b Find the distance around the crop circle if you walked on the outermost circle.

c Find the radius of the inner circle if its area is half the area of the whole crop circle.

9 Stonehenge in England is one of the most mysterious human landscapes ever constructed. Nobody is exactly sure how the stones (some of which weighed as much as 23 metric tonnes) were moved, much less lifted, to create the structure. An outer circle of sarsen stones surrounds a smaller horseshoe of bluestones, with the structure seeming to point in the direction of

the sunrise during the summer solstice. The stones are arranged in a circle with the diameter of 33 metres. It is believed that Stonehenge was built on sacred ground and that among other things it was a burial site. Despite the many theories, scientists are still trying to unravel the mystery of Stonehenge.

The sarsen stones in the outer circle can be modeled by rectangular prisms, with a base measuring 2.1 m by 1.1 m and a height of 4.1 m. It is thought that there were 30 such stones in the outer circle.

a What area does Stonehenge cover?

b Find the total volume and surface area of the 30 sarsen stones.

One of the most famous buildings in New York City is the Flatiron building shown here. It was designed to make the most of a small triangular lot, but because of its shape as a triangular prism with a height of 285 feet, most people initially thought it would eventually fall down. Over 115 years later, it stands as one of the most iconic structures in the Big Apple. The base of the building is a right-angled triangle with a height of 173 feet and a base measuring 87 feet.

a The building was designed to house offices for the company that built it. Find the maximum volume of office space that could be expected.

b **Explain** why the actual available space for offices is less than the value you calculated in **a**.

Because all sides of the building face streets, occupants have offices that are very well lit. The middle 12 stories (with a total height of 156 feet) of the building have windows that measure 7 feet by 2 feet. The length of the long side of the building (triangle) is 193.6 m.

c What percentage of the surface area of the middle 12 stories is covered in windows if there are a total of 528 windows in these 12 stories?

11 The Marianas Trench is the deepest point on Earth, located in the Pacific Ocean, east of the Philippines. The trench was formed by the process of *subduction*, when one tectonic plate dives under another (in this case, the Pacific plate dives under the Philippine plate). The trench can be modeled by a triangular prism with a base width of 69 km and a base height of 11 km (the deepest part of the ocean). The trench has a length of 2500 km.

a Sketch a diagram of the trench and indicate its dimensions.

b Find the volume of water contained in the trench if the sides were perfectly smooth.

12 With the invention of the elevator in 1853, it became possible to imagine creating tall buildings where large numbers of people could work or live. City centers were created with these 'skyscrapers', often in the shape of rectangular prisms. Regulations on these buildings have evolved so that they now require a specific amount of fresh air, q, to be supplied to the building, where q is measured in m^3/h.

The formula for calculating this fresh air supply is given by:

$q = nV$, where n is the number of air changes per hour and V is the volume of the building measured in m^3.

a If the number of air changes per hour needs to be 4, find the fresh air supply for a building in the shape of a rectangular prism with a length of 40 m, a width of 45 m and a height of 250 m.

b The length of an office building is 30 m and its width is 50 m. Find the maximum height of the building if it needs 4 air changes per hour and the ventilation system will provide fresh air at a rate of 19 500 m^3/h.

ATL1 **13** In 1988, scientists discovered a hexagonal storm pattern on the north pole of the planet Saturn. The sides of the hexagon measure 13 800 km while its apothem measures approximately 12 000 km. Readings from NASA's Cassini mission suggest that the structure has a depth of roughly 100 km.

At the center of the hexagon storm is a swirling circular storm (called a cyclone) that measures approximately 3 000 km in diameter.

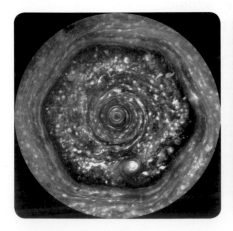

a Find the area of the hexagonal storm at the north pole. Show your working.

b Find the percentage of the hexagonal storm's area that is occupied by the central circular cyclone. **Show** your working.

c Find the volume of the hexagonal prism formed by the storm structure. **Show** your working.

d Calculate the percentage of the hexagonal prism's volume that is occupied by the central cylindrical structure.

14 On average, Los Angeles, California, receives less than 15 inches of rain every year, but it is home to approximately 4 million inhabitants. Supplying water to this many people requires transporting it from far away through the use of aqueducts, like those shown below.

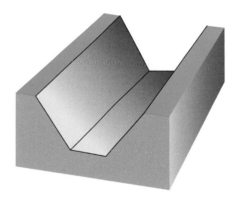

These channels, which extend over 80 km of land, were built with a bottom width of 25 meters while the width at the top is 45 meters.

a If the vertical height of the aqueduct is 5 m, what is the maximum volume of water the aqueduct could hold over a length of 1 km?

b Explain why you think this shape was used instead of a rectangular prism.

15 For what radius value will the circumference of a circle be numerically equal to its area?

The subway system in Montreal, Canada, called 'the metro', opened in 1966 with 26 stations on three separate routes (also called 'lines'). Building the tunnels for the metro involved removing an incredible amount of dirt and rock that was then used to enlarge and even create islands, such as Ile Notre Dame, in the St. Lawrence River near the city. In this task, you will make a proposal to the city planner for the creation of two islands with the material excavated from the construction of a subway system.

In order to make the 50 km of tunnels for the subway, a tunnel boring machine (TBM) will be used that creates rectangular tunnels with sides of 8 m and a height of 6 m. Within the system, there will be 40 stations, each designed in the shape of a rectangular prism measuring 180 m in length, 14 m in height and 20 m in width.

You will create two design proposals, each consisting of two islands. The islands will sit in a river that is 30 m deep, and they will need to be flat, with at least 10 m of land above the level of the water. In order to retain the dirt and give stability to the island, concrete blocks will be placed in the shape of each island and then filled in with the excavated dirt. These blocks will extend all the way from the bottom of the river to the ground level of the island.

This proposal is to be done individually and submitted in a written report.

The proposal

- You must propose two designs for a pair of islands that use as much of the excavated dirt as possible (within 10%).
- You must suggest names for the islands and propose uses for them (e.g. Ile Notre Dame has a park with a public beach and was used for a World's Fair).

Requirements

- Find the volume of dirt and rock that will be excavated for the subway system. Create diagrams and show all of your working.
- Create diagrams of your two proposals.
- Calculate the volume of dirt required to make the islands.
- Show that the total volume of dirt required to build each one is equal/almost equal (within 10%) to the volume of dirt excavated.
- Find the total area that the concrete covers.

Your teacher may want to showcase everyone's ideas for the islands and ask you to draw both designs on one Google slide so all designs can be compiled into a presentation. You can use Tinkercad or Google Sketchup to create to-scale blueprints of your islands.

Reflection

- As stated in the question, these islands were constructed in Montreal. Research these islands. What do you think of the islands created?
- Which exhibits more order, natural or human landscapes? Use 2 examples of both to support your response.

6 Rates

Systems of currency and measurement are ideal places to look at the application of rates, as you will see in this unit. However, rates are an integral part of much of what you see around you. Your study of rates could take you back in history, to a time when computers were just being introduced, or it could help you to understand how the world around us continues to evolve.

Scientific and technical innovation

The space race

After World War II, the United States and the Soviet Union were embattled in what was known as the Cold War. In the 1950s, however, that battle traveled to outer space as the two powers competed in "the space race".

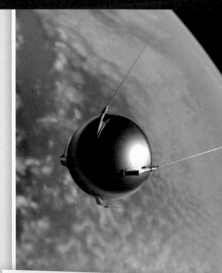

In 1957, the Soviet Union launched Sputnik, the first satellite to orbit Earth. One year later, the United States launched its own satellite, Explorer I, and President Eisenhower created NASA (the National Aeronautics and Space Administration). With the technology of the time, how fast could the rockets carrying the satellites travel? How much was each country willing to spend on these space projects? Who was able to spend more and did that impact the results?

In 1959, the Soviets landed a probe on the moon and, just two years later, one of their astronauts was the first to orbit Earth. The United States continued to lag behind their competitors until landing the first humans on the moon in 1969.

How long would the astronauts' oxygen last on the moon? What were the effects of the moon having one-sixth the gravity of Earth? These questions can be answered using rates.

Evolution and adaptation

Whether because of a loss of habitat or the introduction of a new species, animals and plants evolve and adapt. How fast do such changes take place? How serious does the new threat have to be before adaptation takes place? These rates of change are important pieces of data which help us to understand how one species may affect another.

On the Samoan islands, the blue moon butterfly has demonstrated rapid changes because of a parasite that was destroying the male embryos. Within 5 years, the male population had rebounded from being just 1% of the population to nearly 50%. It was not that the parasite had vanished, but a mutation allowed male butterflies to survive the parasite and, therefore, future generations of males were born with the new characteristic.

Did you know that crabs are invading New England? The Asian shore crab is an invasive species found on the east coast of the United States, though scientists are not sure exactly how they got there. The crabs tend to feed on blue mussels, which are native to the region. However, blue mussels are adapting to these new predators by developing a thicker shell. Only those mussels with the thicker shell are surviving, thereby changing the species, but only in regions where the Asian shore crab is prevalent.

The rate at which the native population was decreasing as well as the rate at which it increased after the mutation is of importance to researchers.

6 Rates

Interconnectedness of human-made systems

Related concepts: Equivalence and Measurement

Global context:

In this unit, you will explore different systems used to measure everything from distance to temperature to monetary value. As part of the global context **globalization and sustainability**, the interconnectedness of human-made systems is just one example of how, despite differences in our development, societies are all undeniably connected. How we negotiate between these systems (where there are differences) is the focus of this unit.

Statement of Inquiry:

Establishing relationships of equivalence between measurements illustrates the interconnectedness of human-made systems.

Objectives

- Defining rate and unit rate
- Converting between different units of measurement and between different currencies
- Defining and recognizing a constant rate of change
- Applying mathematical strategies to solve problems using rates and unit rates

Inquiry questions

F
What is a rate?
What does it mean to be equivalent?

C
How are relationships of equivalence established?

D
How can 'different' still be equivalent?
Do our different systems hinder our interconnectedness?"

ATL1 Creative-thinking skills

Make guesses, ask 'what if' questions and generate testable hypotheses

ATL2 Critical-thinking skills

Draw reasonable conclusions and generalizations

You should already know how to:

- multiply decimal numbers
- perform mathematical operations with fractions
- solve problems with percentages
- solve proportions
- plot points on a Cartesian plane

Introducing rates

What is it like to live somewhere else? If you had to move to a different continent, would life be the same? Do jobs have the same salary from one country to the next? If you moved from New Zealand to India, would you be willing to pay 1 300 000 rupees for a hamburger? Would you or your family know what it means when you see that the speed limit on the highway is 30? If you were offered a summer job that paid 3 Kyat per day, would you be excited or horrified?

In this unit, you will look at how to navigate different systems of measurement and currency. Being different isn't necessarily better or worse, and understanding how to convert between systems will help you determine just how equivalent measurements in each system may be.

ATL1 Reflect and discuss 1

- Some countries have currencies where values are always in thousands or millions. Why do you think they chose to do this?

- The euro is used in approximately 25 countries. Why do you think these countries adopted this common currency?

- Why do you think countries around the world use different units for the same measurement? Why aren't units used universally?

What is a rate?

Whereas a ratio compares amounts measured in the same unit, a *rate* is a comparison of two quantities measured in *different* units.

Suppose you are traveling by car in Iceland with your parents and you see a sign like the one here with the number 90 on it. You ask your parents what it means and they reply that it's the speed limit, 90 kilometers per hour (90 km/h). This is the *rate* of speed, where the quantities are measured in kilometers (distance) and hours (time).

As you travel, you will need to fill your car with gas or petrol. At the petrol station you notice another sign. The price for one grade of petrol (SUPER) is 1.69 euros per liter (1.69 €/L). In this rate, the units are euro (money) and liters (petrol).

When you stop for snacks, you may see a deal for a four-pack of water for €3.50.

'€3.50 for 4 bottles' of water is also a rate, and so is '4 bottles for €3.50'.

Any time you are comparing quantities that are measured in different units, you are looking at a rate.

Activity 1 – Daily rates

Your teacher will provide you with flyers or newspapers that you will use for research.

1 Read the flyers or newspapers and write down ten examples of rates.

2 What are the units of each quantity? What are these units measuring?

3 Are higher values or lower values better for each rate? Explain.

4 What is the most common type of rate that you see in the flyer/newspaper that you have?

Converting measurements

Humans have been creating and changing systems of measurement for centuries. In England, English units were first used in the 15th century. They were replaced by the imperial system of measurement in the 1824 British Weights and Measures Act, and these were eventually replaced by the metric system. The United States developed their own system of measurement units from the English units, and still use it today.

How do you compare measurements in one system with measurements in the other? For example, at 7 feet 7 inches tall, Manute Bol was the tallest basketball player ever in the United

States National Basketball Association. Suleiman Ali Nashnush was the tallest basketball player on the Libyan national team at 246 cm. Who was taller?

Activity 2 – Unit conversions

The US customary units for length are the following: inch, foot, yard and mile. Their equivalents are given below:

$$1 \text{ foot} = 12 \text{ inches} \qquad 3 \text{ feet} = 1 \text{ yard} \qquad 5280 \text{ feet} = 1 \text{ mile}$$

1 Write each conversion as a rate, including the units.

2 Copy and fill in a table like this one.

Inches	Feet
12	1
24	
36	
60	
120	

ATL2

3 Does this conversion relationship represent a proportion? Do all of them? Explain.

4 How many feet is 96 inches? Solve this problem using a proportion.

5 Convert 3.5 feet to inches using a proportion.

Converting from the US customary units to metric requires equivalencies. The most common ones are given below:

$$2.54 \text{ cm} = 1 \text{ meter} \qquad 1 \text{ mile} = 1.61 \text{ km (approximately)}$$

Sometimes it takes more than one calculation to convert units. Another way to perform conversions is to multiply by the *conversion factor(s)*. Suppose you wanted to convert 2 km to inches.

6 Fill in the following with the appropriate conversions.

$$2 \text{km} \times \frac{1 \text{ mile}}{\square \text{ km}} \times \frac{\square \text{ feet}}{\square \text{ mile}} \times \frac{\square \text{ inches}}{\square \text{ feet}}$$

7 Simplify this expression and write your answer as a decimal rounded to 2 decimal places.

> Notice how all of the units simplify to just leave the desired unit of inches in the numerator:
>
> $$2 \cancel{\text{km}} \times \frac{1 \cancel{\text{mile}}}{\square \cancel{\text{km}}} \times \frac{\square \cancel{\text{feet}}}{\square \cancel{\text{mile}}} \times \frac{\square \text{ inches}}{\square \cancel{\text{feet}}}$$

8 Convert 3 yards to meters using this same process. Round your answer to the nearest hundredth.

Example 1

Q Who was taller: Manute Bol, who was 7 feet 7 inches tall, or Suleiman Ali Nashnush, who was 246 cm tall?

A Manute Bol:

$$7 \text{ feet} \times \frac{12 \text{ inches}}{1 \text{ foot}} = 84 \text{ inches}$$

> Convert Bol's imperial measurement to metric.

> First convert the feet to inches using the rate $\frac{12 \text{ inches}}{1 \text{ foot}}$.
> Notice how the units of foot/feet simplify.

$$84 \text{ inches} + 7 \text{ inches} = 91 \text{ inches}$$

> Find Bol's total height in inches.

$$91 \text{ inches} \times \frac{2.54 \text{ cm}}{1 \text{ inch}} = 231.14 \text{ cm}$$

> Convert from inches to centimeters using the rate $\frac{2.54 \text{ cm}}{1 \text{ inch}}$.
> Notice how the units of inches/inch simplify.

Manute Bol was 231.14 cm tall, compared with Suleiman Ali Nashnush at 246 cm tall.

Suleiman Ali Nashnush was taller.

Did you know?

In 1628, the warship Vasa set sail from Sweden only to capsize less than 1300 meters from shore. The ship lacked stability, in part because different teams of workmen used different systems of measurement. Archaeologists have found four rulers that were used when building the ship. Two of them were calibrated using Swedish feet, which are equivalent to 12 inches. The other two rulers, however, were calibrated using Amsterdam feet, which are just 11 inches long!

Practice 1

1 Convert the following measurements to the indicated unit. Show your working and round all of your answers to the nearest tenth.

a 3 km = _____ miles   **b** 60 cm = _____ inches   **c** 2 miles = _____ yards

d 4 yards = _____ cm   **e** 100 m = _____ miles   **f** 5 km = _____ feet

2 Another common place where you find different systems of measurement is in recipes. Not only do you have teaspoons, tablespoons and cups, but you can also find recipes where measures are given in milliliters (mL). The most common conversions for liquid measures are given below.

       1 tablespoon = 3 teaspoons   ½ cup = 8 tablespoons   1 cup = 240 mL

Convert each of the following. Show your working.

a ¾ cup = _____ mL   **b** 700 mL = _____ tablespoons

c 4 tablespoons = _____ mL   **d** 10 teaspoons = _____ mL

e 100 mL = _____ tablespoons   **f** ⅛ cup = _____ teaspoons

3 The 100-yard dash was a common event in the Commonwealth Games as well as being part of the decathlon in the Olympics. It was eventually replaced by the 100-meter sprint.

a Find the difference in race lengths in centimeters. Show your working.

b Linford Christie ran the 100-yard dash in 9.3 seconds. Find how long it would have taken him, at the same rate, to run the 100 m sprint. Show your working.

4 The *Gimli Glider* is the name given to an Air Canada plane that was forced to land in Gimli, Manitoba, when it ran out of fuel. The new Boeing 767 jet was calibrated using the metric system and Canada was just changing its measurement system to metric. However, the fuel calculations were done with imperial measurements.

For this problem, use: 1 liter of fuel has a mass of 0.803 kg

a The amount of fuel on the jet before take-off was 7682 liters. Convert this to kilograms. Show your working.

b The plane required 22 300 kg of fuel for the flight. Find how many additional kilograms of fuel would be necessary for the flight.

c Convert this mass of fuel back to liters to find how many additional liters of fuel were necessary for the trip. Show your working.

1 liter of fuel has a mass of 1.77 pounds (imperial measure). The ground crew used this conversion factor, thinking they were converting to kilograms.

▶ Continued on next page

d Perform steps **a** through **c** using this erroneous conversion factor to find how much additional fuel was *actually* put in the plane's tanks.

e How much fuel was missing because of the use of the wrong conversion factor?

Converting currencies

Not only do countries around the world use different units to measure, but they use different currencies. For example, in South Africa the currency is the rand (R) whereas in Canada it is the Canadian dollar ($ Cdn). You need to be able to convert between different currencies, because one country's currency is not always accepted in another. For example, you cannot walk into a supermarket in Australia and buy a liter of milk with Indian rupees. The amount of one currency you get in exchange for a different currency is called an *exchange rate*.

Activity 3 - Exchanging money

You grew up in Ireland and are about to work for two months during the summer in Thailand. You are offered a salary of 30 000 Thai bhat per month. You are hoping to make 1700 euro for the summer. Given the exchange rate of **1 euro = 39.86 Thai baht**, will you be able to make your goal amount?

1 Copy and fill in a table like this one.

Salary (euro)	Salary (baht)
1	39.86
10	398.6
100	
1000	
10 000	
30 000	

2 Is the relationship between the two currencies a proportional one? Explain.

3 Solve this problem using a proportion. Show your working.

4 Convert 1700 euro into Thai baht using this exchange rate. Show your working.

ATL2

5 Describe a way to use an exchange rate to convert from one currency to another.

▶ Continued on next page

6 Research online for the current exchange rate for 1 Thai baht to euro, as well as the rate for 1 euro to Thai baht. Which exchange rate would you prefer to use: the one in this Activity or the current one? Explain.

7 Using the exchange rates you found for the question above, convert 100 euro to Thai baht and back to euro again. Are these exchange rates equivalent? How do you know?

ATL1 **Reflect and discuss 2**

- Exchange rates posted online are generally not the same as the ones you see at a currency exchange office. Why do you think this is?

- Exchange rates change daily. What do you think affects the value of a country's currency? Give an example.

- Are conversion relationships (measurements and currency) proportional relationships? Explain.

- Create your own 'what if…' question related to exchange rates (that does not require calculations) and discuss with a peer.

WEB LINK

Go to: http://mrnussbaum.com/billions/ and play **"Burnside's Billions"**
Get ready to buy and sell famous world landmarks! With a fortune in US dollars, you will use currency conversions to make an offer for a landmark in the national currency where the landmark is located. You can buy and then sell for a profit, but you may also lose money! Use a spreadsheet for best results as you become a "landmark realtor".

Practice 2

1 Using the sample exchange rates in this table, convert each of the currencies below it. Show your working.

1 Malaysian ringgit = 5411 Vietnamese dong	1 Australian dollar = 51.50 Indian rupees
1 British pound = 1.32 US dollars	1 Argentine peso = 0.048 euros
1 euro = 1.14 Swiss francs	1 Canadian dollar = 39.12 Dominican pesos

a 45 Canadian dollars to Dominican pesos

b 120 000 Indian rupees to Australian dollars

c 55 British pounds to US dollars

d 955 Argentine pesos to euros

e 78 Swiss francs to euros

f 356 000 Vietnamese dong to Malaysian ringgit

▶ Continued on next page

2 Given the following scenarios, use rates to help solve each problem.

a You are from Turkey and going on vacation to Hong Kong. In Turkey, the latest smartphone costs 4092 Turkish lira. However, you see the same smartphone advertised in Hong Kong for 6412 Hong Kong dollars. Given that 1 Hong Kong dollar = 0.44 Turkish lira, is it a better deal to buy it in Turkey or Hong Kong? Justify your answer with mathematical calculations.

b You currently live in Montana, USA, and are looking for a summer job. The minimum wage is $7.25 US. The other option is that you can live with your aunt and uncle in New Zealand for the summer. The minimum wage there is $15.12 New Zealand. Given that $1 US = $1.38 New Zealand, where would you make more money (assuming you work the same amount of hours)?

3 Before currency was ever used, people would use a barter system to acquire goods. In a barter system, each person involved trades something that they have for something they desire. The ancient Egyptians used to trade sheep for other goods. For example, two sheep might have been traded for 25 loaves of bread.

a Write this information as a rate.

b Use this exchange rate to find how many loaves of bread a person could get for six sheep. Show your working.

c Use this exchange rate to find how many sheep a person could get for 40 loaves of bread. Show your working.

d What do you think were some of the issues with a barter system?

4 In the early history of Canada, European settlers would trade a variety of items for beaver skins. First Nations traders could get 35 knives in exchange for two beaver skins.

a Write this information as a rate.

b Use this exchange rate to find out how many knives could be obtained for ten beaver skins.

c Use this exchange rate to find out how many beaver skins could be obtained for 50 knives.

d Suppose First Nations traders could get six copper tools for ten knives. How many skins would they need to trade in order to be able to afford 42 copper tools? Show your working.

Unit rates

Rates can compare any quantities in different units. However, when the second quantity is 1, this is called a *unit rate*. Some rates, such as exchange rates, are generally given as a unit rate since they are easier to use, for example: 0.13 euro / 1 Chinese yuan. However, how do you convert a rate into a unit rate?

ATL2 Investigation 1 – Unit rates

1 On a trip, a family drove 270 kilometers in 3 hours.

   a Find how many kilometers were traveled in one hour. Express this as a rate per hour.

   b Explain the process you used to find the unit rate.

2 You exchange 30 000 South Korean won for 22 euros.

   a Find how many South Korean won are equivalent to 1 euro.

   b Find how many euros are equivalent to 1 South Korean won.

   c Compare the process you used to find the unit rate to the one you wrote in step **1**.

3 Generalize a rule to find any unit rate.

4 Verify your rule for the following rates.

   a $44.00 for 40 liters of gasoline.

   b A six-pack of yogurt costs 1.80 euros.

5 Justify why your rule works.

Reflect and discuss 3

- How is a unit rate different than a ratio represented in simplified form?

- Explain how a unit rate is helpful.

- What is the general rule to change any ratio **a : b** to a unit rate?

Unit rates can be especially helpful when trying to determine which choice of product is more economical. In this case, you would look at the *unit price*, the price for one unit of the product (e.g. one paper clip, one gram, one liter, etc.).

Example 2

Q In one store, you find a 600 mL bottle of soda for AU$3.36 (Australian dollars). In another store, you find a 1.25 L bottle of the same soda for AU$6.75. Which is the better buy?

A Store 1: $3.36 : 600   Store 2: $6.75 : 1250

> Write the ratio of price: amount. Make sure each amount is in the same unit (milliliters).
> 1.25 L = 1250 mL

Store 1: $\dfrac{\$3.36}{600} : \dfrac{600}{600}$ Store 2: $\dfrac{\$6.75}{1250} : \dfrac{1250}{1250}$

> In order to find a unit prices, divide each ratio by the amount of soda.

Store 1: $0.0056 : 1

> Divide to produce the unit price. This tells you how much 1 mL costs.

Store 2: $0.0054 : 1

The better buy is in Store 2, where the unit price for 1 mL is lower.

Reflect and discuss 4

- Give three examples of where you see unit prices instead of the total price for an amount of items.

- Describe how you would have found a unit rate per liter in Example 2. Would using this unit rate affect the outcome of which product is the better buy? Explain.

ATL2

- If you wrote the above ratios as 600 : $3.36 and 1250 : $6.75 instead, and then converted them to a unit rate, what would the unit rate tell you? How would you know which is the better buy? Explain.

Activity 4 – Smart shopper

1 Go on a field trip to a local store (or look online) and find an item that comes it at least three different sizes (e.g. milk, juice, flour, rice).

2 Set up a table to record the data for the different sizes of the same item. You will need to record price and volume so you can calculate the unit price.

3 Which size is the better value? Why?

4 Are there times when you would *not* choose the item that is the best value? Explain.

Unit rates can also be helpful in solving problems and making other kinds of decisions.

Pairs

Activity 5 – Narrator and scribe

Using unit rates, you will compare two situations and decide which one is the better option.

1 Your teacher will split the class in half. You will be given either the letter A or B. To begin, you may select any of the four scenarios in the table that follows. (You will eventually do all of them.) Pairs of students are formed with each pair having a Person A and a Person B.

2 Person A is the 'narrator' for problem A. Person B is the 'scribe'. The narrator tells the scribe how to solve the problem, and the scribe writes down the steps until it is solved.

3 Once you finish with problem A, switch roles. Person B is now the narrator for problem B, and Person A is the scribe.

> Within this activity the scribe can ask questions of the narrator. It is important that pairs work together to understand how to solve each problem.

4 After completing parts A and B in the scenario, answer the question at the end of problem B.

5 When you complete your first scenario, move on to another scenario. Repeat until all scenarios are completed.

Scenario 1: find the unit rate in each case

| (A) You are traveling from Ireland to Croatia on vacation. You get to the airport and exchange 120 euros and receive 852 Croatian kuna. | (B) You are traveling from Ireland to Croatia on vacation. You wait to exchange money until you get into town, where you exchange 350 euros and you receive 2600 Croatian kuna. |

Which exchange rate was better?

Scenario 2: find the unit rate in each case

| (A) You can walk $\frac{1}{2}$ a kilometer in $\frac{1}{4}$ of an hour. | (B) You can walk 5 kilometers in 2 hours. |

Who walks faster, Person A or B?

Scenario 3: find the unit rate in each case

| (A) A company in Russia offers to pay you 17 000 rubles for 9 hours of work. | (B) An opposing company in Russia offers to pay 26 000 rubles for 14 hours of work. |

Which company pays more?

▶ Continued on next page

Scenario 4: find the unit rate in each case

(A) You get a job delivering food in America. Beyond getting paid regularly, the company will pay you $\frac{1}{4}$ of a dollar every $\frac{1}{8}$ of a mile.

(B) A different company offers the same regular pay but pays $\frac{1}{2}$ a dollar every $\frac{3}{4}$ of a mile.

Which company pays better?

ATL1 ## Reflect and discuss 5

- The differences between some unit rates can be very small. Why should you care about such a small difference? Give an example to support your claim.

- In your grocery store, are the prices of goods given as unit rates? How would you use this information when you shop?

- Create a 'what if…?' question related to rates. Discuss your question with a peer.

Practice 3

1 Find the unit rate for the following ratios. If it is already a unit rate, write 'unit rate'.

a 6 : 2	**b** 1 : 5	**c** 5 : 1
d 165 : 55	**e** 225 : 15	**f** 7 : 14
g 2000 : 1000	**h** 408 : 51	**i** 66 : 11
j 81 : 135	**k** 8246 : 434	**l** 98 : 490

2 Find the unit rate for each of the following.

a 12 meters per 8 seconds	**b** −20 degrees per 5 hours
c 8 euros per 4 bags	**d** 180 Thai baht for 5 pairs of sunglasses
e 9 movie tickets for 72 British pounds	**f** 100 kilometers per 8 hours
g 60 cm of snow in 12 hours	**h** 30 staff members for 5 trains

3 Mary ran 3 miles in 32 minutes, and Kiara ran 4 miles in 41 minutes. Using unit rates, who ran faster?

4 Bhim exchanges 50 000 pesos for 2200 euros. Find how many pesos can be exchanged for 1 euro.

▶ Continued on next page

5 Where possible, find the unit rate for the following ratios.

a 30 : 6   b 3 : 2   c 6 : 10

d 1 : 8   e 5 : 1   f 21 : 6

6 Find which choice is the better buy. Show all of your working.

a

29.27 R for 6 eggs   170 R for 36 eggs

b

2 L milk for $3.95 CDN   473 mL milk for $0.93 CDN

c

454 g rice for €0.60   9.07 kg rice for €10

d

3 bananas for 1.10 R   7 bananas for 2.75 R

▶ Continued on next page

7 Find the unit rate, then compare the following and decide which is the better option.

a 50 000 Chilean pesos for 1200 British pounds, or 1350 Chilean pesos for 40 British pounds

b 600 Chinese yuan for 1700 Mexican pesos, or 1400 Chinese yuan for 4300 Mexican pesos

c 7900 Japanese yen for 84.50 Australian dollars, or 12 500 Japanese yen for 144.50 Australian dollars

d 9800 Mauritian rupees for 3900 South African rand, or 1400 Mauritian rupees for 635 South African rand

e 8700 Albanian lek for 55 000 Chilean pesos, or 1700 Albanian lek for 11 000 Chilean pesos

8 You have been shopping around for the best exchange rate for your trip to Hungary. You found two options that seem reasonable.

- Option 1: 261 Hungarian forint for 1 US dollar

- Option 2: 270 Hungarian forint for 1 US dollar

Samantha is exchanging 800 000 Hungarian forint for US dollars.

a How many US dollars would she get with the exchange rate at option 1?

b How many US dollars would she get with the exchange rate at option 2?

c Why is it important to research where to get the best exchange rate?

Formative assessment

criterion **C, D**

You are going to plan a celebration for your class! You will shop online for ten products that you know you will need, and then find where to get the best buy.

1 Write down the ten different items that you would like to buy for the celebration. Half of them should be food items and half will be other items (e.g. party supplies). Be sure to indicate how many and what size, if appropriate.

2 Look online or in flyers for possible options for each. You must compare at least two different size options for each item. Use unit prices to compare the items and indicate which is the better buy.

3 Record all of the options in a table and present a budget for the list of options selected.

4 Select a country in a different IB region than yours. Research the price of the ten items in that country and calculate the total in their currency.

▶ Continued on next page

5 Research the exchange rate for that country and convert the cost of the items to your country's currency. Where are the items less expensive?

Reflection

- When do you think it is important to comparison shop for the better buy? Explain using clear examples.

- There are companies that import goods from overseas and then sell them. When do you think this is an effective business practice? Explain.

Problem-solving with rates

As well as currencies, measurements and prices, many other kinds of problems can be solved through the use of rates.

Example 3

Q There are seven painters who are painting a building. Each of them puts in 80 hours to do the job. How long would it have taken just two painters to do the same job, if they all work at the same rate?

A $7 \text{ painters} \times \dfrac{80 \text{ hours}}{1 \text{ painter}} = 560 \text{ hours}$

> Find the total number of hours it takes to do the job.

> Notice how the units of 'painter' simplify, leaving the unit of 'hours'.

$560 \text{ hours} \div 2 \text{ painters} = 280 \text{ hours per painter}$

> Find out how many hours the new number of painters will need to do the job.

It will take two painters 280 hours each to paint the same building.

Practice 4

1 An animal travels 3 km in 8 hours. Find how long will it take the same animal to travel 300 m, assuming it travels at a constant speed. Show your working.

2 Jamal read a 300-page book in 12 hours. At the same rate of reading, how long would it take him to read a 400-page book? Show your working.

3 Water from a shower flows at a rate of 8 liters per minute. Find the amount of water (in liters) that a family of five will use in a day, if the average shower lasts 7 minutes per person.

▶ Continued on next page

4 The average person's heart beats 70 times per minute. It can be shown that the average person's heart beats 2.57 billion times in a lifetime.

   a According to this information, how long is the average life span?

   b What assumptions have to be made to solve this problem? How do they affect your answer to part **a**? Explain.

5 Working at the same time, it takes five people 20 minutes to build five models. How long would it take seven people to build seven models? Show your working.

6 It took nine students 35 hours to build the set for a musical theater production.

   a How long would it have taken 12 students to build the same set? Show your working.

   b If the set needed to be built in less than one day, how many people would that have required? Show your working.

7 You and three friends are on wilderness trip and have exactly enough food for 5 days. There are 60 set meals and each person eats 3 meals per day.

But on the first morning of the trip you meet a person on the trail who has been injured, and you call for a medical unit to come get him. The area is so remote that you are told that the earliest a helicopter can come to the rescue is 3 days.

Will there be enough food for all five of you until the helicopter arrives? Justify your answer.

Exploring rate of change

While rates compare quantities measured in different units, sometimes those quantities change over time. How they change is an important concept in mathematics, called the *rate of change*. Understanding the rate at which a quantity changes over time allows you to make predictions which may save you money or even save a life!

ATL1 **Reflect and discuss 6**

- What do you think a 'rate of change' is? Give an example.

- What do you think a constant rate of change is?

Activity 6 – Earning money

1 You accept a job in Switzerland for the summer working at a summer camp. You are offered 4000 Swiss francs for 8 weeks of work. Use a unit rate to find how much you make per week.

2 Copy this table and fill it in, showing how much you will have made in total at the end of every week.

Week	Total money earned (Swiss francs)
1	
2	
3	
4	
5	
6	
7	
8	4000

3 Plot your results on a grid like the one below.

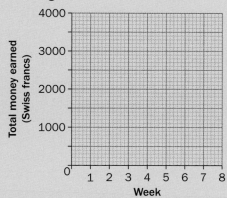

4 Explain what these two points mean in the context of this problem:

   a (0, 0)   **b** (1, 500)

5 How would you describe your graph?

6 Why does the graph look like this? Explain.

Activity 6 is an example of a *constant rate of change*. Quantities can change in many different ways, so what does the graph look like when it is *not* constant?

Activity 7 – Investing money

After earning those 4000 Swiss francs, you want to save the money you have made, so you decide to invest it.

When you invest money, you can choose a wide range of products. Some people put their money in a bank, others in investments. There are few guarantees when you invest money, but in general, the longer you have your money invested, the more money you are likely to make.

Your local bank is offering an amazing deal of 5% interest compounded yearly for the next 6 years. This means that every year the bank will add 5% of your bank balance to the current amount. In this way, as long as you don't remove money from your account, in addition to earning interest on your investment, you will earn interest on your interest!

1 Copy and complete a table like this to see the effects of compound interest on your investment.

Year	Interest calculations	Amount in bank account (Swiss francs)
0		4000
1	4000 + 5% of 4000 = 4000 + 0.05 × 4000 = 4000 + 200	4200
2	4200 + 5% of 4200	4410
3		
4		
5		
6		

2 Plot your results on a grid like the one below.

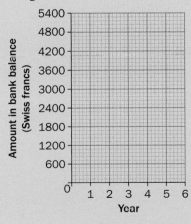

▶ Continued on next page

245

3 Explain what the following points mean in the context of this problem.

    a (0, 4000)   **b** (1, 4200)

4 Do you think this is a constant rate of change? Explain.

5 Explain how this scenario differs from the one in Activity 6.

ATL2 Reflect and discuss 7

- What is a constant rate of change? Explain.

- What does a constant rate of change look like on a graph? Explain.

- How can you tell if a situation or scenario represents a constant rate of change? Explain.

- Does a proportional relationship show a constant rate of change? Explain using an example.

Activity 8 – Taxes around the world

Sales tax is a tax imposed by the government. It is tax charged on the sale of products. Different countries impose different amounts of sales tax.

In Armenia, the currency is the dram and the sales tax is 20%.

1 Copy and complete a table like this one.

Sales price (dram)	Amount of tax (dram)
0	
1	
2	
3	
10	
20	
50	
100	

▶ Continued on next page

2 Plot your results on a grid like this one.

3 Is sales tax a constant rate of change? Explain how you know.

4 Using two of the rows of your table, find the change in the amount of sales tax and the change in sales price. Express this as a rate: (change in tax) / (change in sales price).

5 Repeat step **4** for each pair of successive rows.

6 What do you notice about your results in step **5**?

7 Is this a constant rate of change? Explain.

Example 4

Q While much of the world uses the metric system and measures temperature in degrees Celsius, there are still places where degrees Fahrenheit are preferred.

degrees Celsius	degrees Fahrenheit
−5	23
0	32
5	41
10	50
15	59
20	68

Does the conversion from one system to the other represent a constant rate of change? If so, find the rate of change.

A Yes, this is a constant rate of change.

> Since each 5 degree change in Celsius represents a 9 degree change in the Fahrenheit scale, this is a constant rate of change.

▶ Continued on next page

Using the last two rows:

The change in Celsius is $20 - 15 = 5$.

The change in Fahrenheit is $68 - 59 = 9$.

The rate is $\dfrac{5}{9}$.

Yes, it is a constant rate of change. The rate of change is $\dfrac{5}{9}$.

Use any two rows and find the change in Celsius and the change in Fahrenheit.

Express as a rate.

Reflect and discuss 8

- Does it matter which quantity comes first in the rate? Explain.

- How would you recognize a rate of change that is not constant? Explain.

Practice 5

1 Given the graphs below, decide which graphs have a constant rate of change and which do not.

a 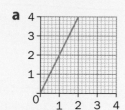 **b** 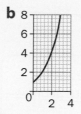 **c** 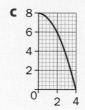 **d**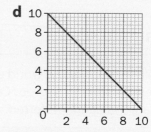

2 Given the situations below, decide which represent a constant rate of change and which do not.

a On vacation you give yourself a daily budget. You spend 200 euros per day for 14 days.

b You are running a race. You start by going 1 km every 6 minutes. After 2 km you speed up and start running 1 km every 5 minutes for the rest of the race.

c You are spending a summer traveling and you want to see as many cities as possible. You decide to visit one city per day for the 10 days you are on vacation.

d You are saving money for a car. The first year you save 200 Swiss francs per month. The next year you save 400 Swiss francs per month. The third year you save 600 Swiss francs per month.

▶ Continued on next page

3 For each of the following tables, identify whether there is a constant rate of change. If there is, graph the relationship and find the constant rate of change. If not, write 'not a constant rate'.

a

0	0
1	2
2	4
3	6
4	8
5	10

b

6	9
7	10
8	12
9	15
10	16
11	18

c

0	0
1	15
2	30
3	45
4	60
5	75

d

20	60
35	70
45	135
50	100
70	280
90	270

e

10	50
11	55
27	135
33	165
48	240
53	265

f

5	5
10	10
15	20
20	40
25	80
30	160

4 Show that an exchange rate represents a constant rate of change. Use an example and show your working.

5 Israel has a sales tax of 15.5%.

a Create a table with four different sales prices and their corresponding sales tax.

b Is this a constant rate of change? Justify your answer.

c Is this situation a proportional relationship? Explain. If so, write an equation to represent the tax expected (t) on any sales price (p).

6 Energy can be measured in a variety of units. The SI unit is the joule (J), but it can also be measured in calories (cal) or kilocalories (kcal). Energy equivalents are given in the table on the right.

Energy (J)	Energy (kcal)
20 920	5
33 472	8
50 208	12
83 680	20

a Find the rate of change between successive rows in the table.

b Is this a constant rate of change? Explain.

c Is this situation a proportional relationship? Explain. If so, write an equation to determine the energy in kcal (k) knowing the energy given in joules (J).

7 The table represents the amount of money you earn over a summer. The number of weeks you work is represented by x, and y represents the money you make after each week.

Weeks worked (x)	Money earned (y)
0	0
1	250
3	750
8	2 000

a Plot the data using an appropriate scale.

b Is this a constant rate of change? If so, find the rate.

c Is this a proportional relationship? Explain.

d Find an equation relating the two variables.

249

Unit summary

A rate is a comparison of two quantities measured in different units.

Converting between measurements and currencies requires the use of a conversion factor.

When the second quantity in a rate is 1, this is called a unit rate. Exchange rates are generally given as a unit rate since they are easier to use.

The unit price is the price for one unit of a product.

A rate of change is how quickly a quantity changes in relation to changes in another quantity. A rate of change between two quantities can be constant, meaning it is always the same. A change in one quantity brings about a consistent change in the other.

The graph of a relationship with a constant rate of change is a straight line, though it does *not* have to go through the origin (0, 0).

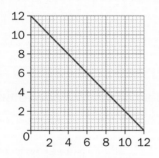
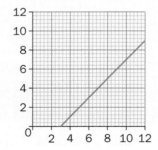

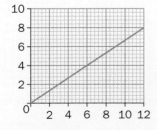
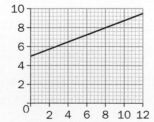

Unit review

criterion **A**

> 🗔 **Launch additional digital resources for this chapter**

Key to Unit review question levels:

| Level 1–2 | Level 3–4 | Level 5–6 | Level 7–8 |

1 Convert the following measurements.

a 300 inches = _____ meters   **b** 26.2 miles = _____ km

c 30 m = _____ yards   **d** 5 tablespoons = _____ liters

e ¾ cup = _____ mL   **f** 500 mL = _____ tablespoons

2 Find the unit rate for the following situations.

a 80 miles per 4 hours

b 120 kilometers per 8 hours

c 20 inches per 30 days

d 15 miles per 40 minutes

3 Determine whether the following situations represent a constant rate of change or not.

a Marta saves 30 euros a day for 20 days.

b Harry runs for 8 kilometers. He runs the first kilometer in 6 minutes, then runs the rest of the distance in 40 minutes.

c While driving her car, Zara uses cruise control and drives 50 miles per hour for 9 hours.

d The snow melts one inch per hour on the first warm day of spring.

4 The fuel economy or gas consumption of a car is given as a rate of distance per volume. A car is advertised as getting 32 miles per gallon in the United States. How would this rate be expressed in Canada (in kilometers per liter) if one gallon is approximately equal to 3.8 liters?

5 The track at a local school is known to be 400 meters long. If Cynthia wants to train for a 3.5-mile race, how many laps around the track should she run? **Show** your working.

6 **Use** this table to convert the currencies listed below it.

1 Brazilian real = 9 Uruguayan pesos	1 Rwandan franc = 0.30 Yemeni rials
1 British pound = 1990 Lebanese pounds	1 Australian dollar = 79 Serbian dinars
1 Dominican peso = 0.05 Belize dollars	1 Barbadian dollar = 0.84 UAE dirhams

a 60 Brazilian reals to Uruguayan pesos

b 900 Serbian dinars to Australian dollars

c 45 British pounds to Lebanese pounds

d 1500 Yemeni rials to Rwandan francs

e 87 Dominican pesos to Belize dollars

f 2000 UAE dirhams to Barbadian dollars

7 **Use** the table above to decide which quantity is higher.

a 700 Brazilian reals or 6500 Uruguayan pesos

b 18900 Lebanese pounds or 12 British pounds

c 150 UAE dirhams or 86 Barbadian dollars

d 13 Rwandan francs or 50 Yemeni rials

8 In 1999, the Institute for Safe Medical Practices reported that a patient was given 0.5 grams of a drug instead of 0.5 *grains*, as was written on the label. The same dose was administered for three straight days before the error was found. If 2 grains are equivalent to 0.13 grams, how many grains did the patient receive over the course of the three days? **Show** your working.

9 For a school fundraiser, 15 students washed 40 cars in 4 hours.

a Find how long it would take 10 students to wash the same 40 cars. Show your working.

b If the goal of the fundraiser was to raise $500 and each driver donated $10 to have their car washed, for how long will those 10 students be washing cars?

10 Joseph is offered a job in Brazil for 8 weeks in the summer, paying 1600 Brazilian reals per week.

a Create a table to represent the money he will make each week for 8 weeks.

b **Construct** a graph using the table you created.

c Create an equation to represent how much money he will make in any given week.

d Describe in words what the point on the graph (1, 1600) means in this situation.

11 In physics, force can be measured in either newtons (N) or pounds (lb).

In 1999, the Mars Climate Orbiter came too close to Mars and burned up in its atmosphere. Part of the problem was that the software used to control the orbiter's propulsion was calculating force in pounds, but a different piece of software using those calculations was calibrated to use newtons. The miscalculation caused the orbiter to be propelled too close to the surface of the planet where it was destroyed.

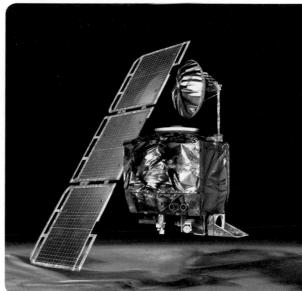

20 newtons is roughly 4.5 pounds of force.

a How many pounds of force is equivalent to 3000 newtons? **Show** your working.

b If the first piece of software calculated a value of 2900 pounds, how many newtons would that have been equivalent to?

12 Rebecca goes on vacation with 2000 dollars. She is on vacation for 20 days. She spends 100 dollars per day.

a Create a table to represent the total amount of money she has spent at the end of each day.

b **Construct** a graph using the table you created.

c Create an equation to represent how much money she will have spent after any given day.

d **Describe** in words what the point on the graph (20, 2000) means in this situation.

Summative assessment

Holiday travels

As you have seen, every country in the world has their form of currency, though the euro has been a base currency for as many as 25 European countries. Exchange rates are used in order to be able to convert from one currency to another.

You will be using this concept to calculate the budget for a holiday you will be taking. Your teacher will let you know if you are working in pairs and going on the trip together, or if you are calculating the budget for a trip for two on your own.

You are going on a roundtrip holiday to five different countries (each with their own currency). You will have only one stop per country and you can travel between these destinations using any mode of transport you wish (car, train, airplane, etc., depending on which is the most appropriate to use).

Currencies

1   Select your five destinations and explain why you have chosen to visit those locations.

2   In order to plan properly, you will have to find the exchange rates between each of the countries you are visiting and your home country. As a class, you will determine how much cash you will budget per day for meals and activities. Keep in mind that most places will accept international credit cards, so large purchases and accommodations can be put on a credit card.

3   Exchange rates between currencies change all the time, so you will need to research the most up-to-date rates from a reputable and reliable website. National bank websites are good for this.

4   Set up a table with the exchange rates you will need and the number of days you plan on spending in each country.

Planning your trip

1   Create an itinerary for your trip, organized by country visited. Your itinerary must include your mode of travel, accommodations, meals and any places of interest that you would like to visit. Print out a map that includes all of the countries you are visiting and trace your round trip on the map.

2   For each leg of your journey, determine the mode of transportation between countries and research the average speed of travel in kilometers per hour. For example, jumbo jets have a cruising speed in excess of 900 km/h. Use Google maps or a similar website to determine the distance between destinations and calculate the approximate time it will take to travel between the cities. Show all of your working. Research the approximate cost of travel between each location in local currency and convert the cost to your home currency.

3 Try to find accommodations of similar standards in each country. Record the cost of accommodations for each country in their currency and convert the cost to your home currency. Show all of your working.

4 Be sure to calculate the cost of meals and excursions/visits in the local currency and in your home currency.

5 Summarize all of these costs (transportation, accommodations, meals, excursions/visits, etc.) in a table, showing the local currency and its equivalent in your home currency. Calculate a budget for the entire trip given how much spending money you will need and all of your other costs. This can be set up in a spreadsheet or by hand, as instructed by your teacher, keeping in mind that all calculations must be shown.

Questions

1 You would like to buy a new backpack either before your trip or from one of the countries you visit. Research the price of the exact same backpack in each of the countries. Is it the same price in all countries? If not, why do you think that is? Where would you consider buying it? Would you make the decision to buy it before you leave or buy it abroad? Justify your decision.

2 Research the minimum wage per hour in your country. How many hours would you have to work in order to save enough money to go on this trip?

3 Create a table, a graph and an equation that represents the relationship between your total earnings and the number of hours worked. Explain why this is a constant rate of change and how this rate of change is represented in the equation, graph and table.

4 Assume you went on your trip and have 10% of the cash left over from each of the five countries. Calculate how much you will get back from your bank in your home currency when you exchange it back home. (Remember: there is a buy and sell rate for each currency.)

Reflect and discuss

- Compare the exchange rate for your home country's currency to the currency of the primary country where you spent the most time on your trip. What is the unit rate? Has the value of your country's currency been decreasing or increasing against the primary visit country? Would this trip have been cheaper or more expensive a year ago?

- Some investors try to make money by watching exchange rates and predicting if a particular currency will get stronger or weaker against other currencies. How might people make money doing this? What are the risks of trying to do this?

- Why do we not all follow and use the same measurement systems? Do our different systems hinder our interconnectedness?

7 Univariate data

Analysing data is an effective way of determining whether or not inequality exists, as you will see in this unit. Collecting data and then representing it in the most effective way is instrumental in helping others to understand your conclusions. However, what would the study of data representation and analysis look like if it were applied to other contexts such as your health or your digital life?

Identities and relationships

Physical and mental health

Your physical and mental wellbeing are important factors in helping you grow up healthy and happy. Many people see them as separate, but there is evidence that they are closely connected.

Your physical health is impacted by the foods you eat, the amount that you sleep, your activity level and a host of other factors. You can collect and analyse data in order to study your diet, your daily routines and even your schoolwork in an attempt to find the optimal combination.

Good mental health helps you to deal better with life's challenges. But what is typical for a teenager? Data can be collected and analysed to determine if you are at risk of experiencing mental health issues or if your habits are negatively affecting your emotions. At the same time, research can be done to see what actions may improve how you feel.

Paying attention to your mental and physical health is an important part of growing up.

Virtual environments and digital life

Your generation is the most connected generation ever. You have not only grown up with the internet, but you will probably have had some kind of connected device nearby your entire life. How has that impacted your adolescence? What advantages has it offered you? Have there been any disadvantages? These are questions that statistics can help to answer.

Screen time refers to the amount of time someone spends in front of a screen, be it a phone, tablet, laptop or television. How much of your life is spent in a digital realm of some kind? How does that compare to other teenagers in your school? What is the average amount of screen time? Does it depend on where you live?

What aspects make up your digital life? How much of your screen time is spent doing homework? How much is spent in virtual environments or playing video games? What about social media? Analysing your own behaviour may give you insights into how to better manage your time and minimize the risks of spending too much time on screen.

7 Univariate data
Accessing equal opportunities

Related concepts: Representation and Justification

Global context:

In this unit, you will use statistics to explore the differences between countries as part of your study of the global context **fairness and development**. Representing data with both graphs and numerical measures will help you decide whether inequality exists, and may even, in some cases, surprise you. Numbers don't lie, right?

Statement of Inquiry:

Different forms of representation can help justify conclusions regarding access to equal opportunities.

Objectives

- Representing data using stem-and-leaf plots and box-and-whisker plots
- Calculating measures of central tendency and measures of dispersion
- Choosing the best method to represent data
- Analysing data and drawing conclusions

Inquiry questions

F How do we represent information?
How do we use numbers to justify conclusions?

C What makes a form of representation effective?
What are the strengths and weaknesses of numerical data?

D How can equality be represented?
How can we use information to instigate change and make a difference?

ATL1 Information literacy skills
Process data and report results

ATL2 Collaboration skills
Practise empathy

You should already know how to:

- categorize data into different types
- read information from a bar graph
- read and construct line, bar and circle graphs

Introducing univariate data

No matter where you live, you need water. It is as vital in all countries, no matter how advanced. In 2014, the city of Flint, Michigan, in the United States changed its water source from Lake Huron to the Flint River, which runs through the city. For decades, untreated waste products from industries, humans and even nature found their way into the river. Because of this, the water was treated with a chemical that eventually corroded the aging water pipes, releasing lead into the city's drinking water.

While drinking water is always tested to insure safe levels of contaminants, samples from sites that were known to have much higher than normal levels of lead and bacteria were not used. Also, two samples that did have high levels of lead were supposedly removed, causing the average levels of lead to be below limits that would have required immediate action.

The situation became even worse, with 90% of people in one community contracting Legionnaire's disease, the cause of which is presumed to be the water. Ten of the victims died, and citizens continued to worry about the children who drank the water and who may suffer the effects of elevated levels of lead in the bloodstream.

In cases like these data analysis can help to draw conclusions on issues that have a dramatic effect on a community. Last year, you learned how to represent data. In this unit, you will learn how to analyse data so that you can draw reasonable conclusions and justify them. Maybe you will be the one to discover a trend that could help a neighborhood, a city or even a country.

Reflect and discuss 1

- How could just two samples change a conclusion about lead levels in a river? Explain using an example.

- How much evidence would you need to be convinced that water from the Flint River is now safe to drink?

- What would it be like to find out that you and your family had been drinking lead-contaminated water for six months?

Representing data: stem-and-leaf plots

Last year, you learned how to use a bar graph, a circle graph and a line graph to represent data. While those representations are excellent ways to display data and even show general trends, they do not necessarily show specific data values, nor are they always effective with numerical data. One representation that *does* allow you to see the actual data values collected is called a *stem-and-leaf plot*.

Example 1

(Q) Lead levels were measured in 20 water sources throughout Europe. The following is a sample of some of the data collected (all measures are in parts per billion).

23 17 9 12 14 24 23 12 8 4

20 15 12 8 28 15 11 10 5 22

(A) Represent the data using a stem-and-leaf plot.

```
0 |
1 |
2 |
```

Create the stem by writing the different tens digits vertically. The highest data value is 28, so the highest 'tens' digit is 2.

```
0 | 4  5  8  8  9
1 | 0  1  2  2  2  4  5  5  7
2 | 0  2  3  3  4  8
```

Write the ones digits next to the appropriate tens digit for each piece of data. Write the data in numerical order. Values that are repeated are also repeated in the stem-and-leaf plot.

Key: 2 | 4 = 24

Write a key so that others know how to interpret the data.

Reflect and discuss 2

- What types of data (categorical, ordinal, discrete, continuous) can be represented with a stem-and-leaf plot? Explain.

- Why would a bar graph or a circle graph be less effective in representing this set of data? Explain.

- What might be some advantages with using stem-and-leaf plots? What might be some disadvantages?

- Why is writing a key an important step in creating a stem-and-leaf plot? Give a specific example showing how people may get confused if you do not create a key.

- How would you describe the given data set? What words/values would you use to represent it? Explain.

Activity 1 – Back-to-back stem and leaf plots

Two sets of data can be represented and compared on a single stem and leaf plot, called a *back-to-back stem and leaf plot*. You use the same "stem" for both data sets and put the leaves for one set of data on one side of the stem, and the leaves for the other set of data on the other side.

The pulse rate of students who have access to physical and health education classes are represented below.

After exercise   Before exercise

(beats per minute)

```
                        |  6 | 0  1  2  2  3  5  6  6
                        |  7 | 0  1  1  2  2  4  4  5  7  7  8
                        |  8 | 0
                        |  9 |
                4   2   | 10 |
          8   4   1   0 | 11 |
  7   6 5 5   2   0   0 | 12 |
      7   3 1 1   0   0 | 13 |
                    2   | 14 |
```

key: 10 | 2 = 102

a How does the back-to-back stem and leaf plot represent two sets of data?

b How does the key help you interpret both sets of data?

c What is the mean, median and mode pulse rate before exercising? Show your working.

d What is the mean, median and mode pulse rate after exercising? Show your working.

e What conclusion(s) can you draw from the stem and leaf plot? Explain how each conclusion is visible in the stem and leaf plot.

f Take your pulse rate right now by counting the number of beats in 30 seconds, and multiplying that amount by 2. Your teacher will gather the data for the class.

g Represent the pulse rate for male and female students on a back-to-back stem and leaf plot.

h How do the pulse rates of the boys and girls in your class compare? How is this represented in the stem and leaf plot?

Practice 1

1 Draw a stem-and-leaf plot for each of the following sets of data. Be sure to create a key for each.

   a 101   113   117   104   112   123   102
      117   118   111   100   120   127   112

   b 8.4 7.6 5.2 8.1 9.0 8.8 7.2 5.1

   c 19.5   19.4   18.2   20.3   21.5   17.1   18.0   20.6   20.5

   d 1113   1120   1117   1119   1128   1126   1135   1111   1126   1125

2 The World Health Organization (WHO) set its standard for the acceptable amount of lead in water at no more than 10 parts per billion. Water quality data (in parts per billion) were collected from several rivers in Asia and displayed in the following stem-and-leaf plot.

```
0 | 8 9
1 | 0 2 3 5 6
2 | 1 1 2 3 7 8 8 9
3 | 0 0 2 4 8 9 9
4 | 5 6 6 8
5 | 2 2 3 3 3 6 8 9
6 | 0 1 5 5 7
```

Key: $2\,|\,1 = 21$

   a How many tests resulted in water quality that met the WHO guideline?

   b China sets its acceptable level of lead at no more than 50 parts per billion. How many tests resulted in water quality that would *not* meet this guideline?

   c How many tests contained lead levels between the United States' acceptable level of 15 parts per billion and China's acceptable level?

   d Are the lead levels represented in the table worrying? Explain.

▶ Continued on next page

3 Represent each of the following sets of data on a back-to-back stem and leaf plot. Be sure to indicate a key for each.

a

Weight at 1 year old (kg)	Weight at 10 years old (kg)
10	27
11	29
12	34
13	31
13	32
12	32
10	33
10	29
9	33

b

Number of students per class (IBA)	Number of students per class (IBAP)
15	16
30	8
26	10
21	9
12	10
25	14
30	15
28	12

c

Male height (cm)	Female height (cm)
152	133
154	148
148	132
161	150
173	125
159	128
150	131
161	138
166	119

4 The number of kilocalories consumed per capita (per person) per day (known as *crop supply*) is a measure of how much food is available in a region. A higher number means that more food is available for each person living there.

Crop supply data were collected for countries around the world, with data for high-income countries on the left and low-income countries on the right.

High income countries   **Low income countries**

```
                          20 | 2  2  5  6  9
                          21 | 1  4  5  5  5
                          22 | 1  4  4  6  9
                          23 | 0  0  1  3  5
                          24 | 0  1  1  2  2
                          25 | 0  0  2
                          26 |
                          27 |
                          28 |
                          29 |
                0  1  4    30 |
       2  2  3  5  7       31 |
       1  3  4  8  9       32 |
       0  0  2  3  6       33 |
       1  2  9  9  9       34 |
             1  1  2       35 |
                3  4       36 |
```

Key: 32 | 1 = 3210

▶ Continued on next page

a Write down the highest and lowest crop supply values for low and high income countries.

b Is there any overlap between crop supply values between the two groups? Explain.

c Write down the most common crop supply value for each group.

d From the data, does it look as though low and high income countries have equal access to food supplies? Explain.

5 Despite advances in treatments and education, HIV continues to be an issue worldwide. Stem-and-leaf plots for the number of new HIV cases in countries categorized as high and low income are shown below.

High-income countries

0	4 5 7
1	0 0 1 4 7
2	0 1 2 4 4 7
3	1 2 5

key: $1\,|\,0 = 1000$

Low-income countries

1	1 3
2	1 2 8
5	5 9
7	1 5
21	0

key: $5\,|\,9 = 59\,000$

a What conclusion(s) can you draw based on the data? Justify your answer.

b What questions do you have about the data that might influence your conclusion(s)?

c Why do you think there is such a difference between the number of new HIV cases in the two regions?

d How effective is a stem-and-leaf plot for representing data so that you can draw conclusions? Explain.

ATL2 e What might it be like to live in a region where HIV infection rates are high? Explain.

Throughout this unit, we have used the terms 'low-income' and 'high-income' to describe categories of countries at different stages of economic development. Many other terms and acronyms are used – developed, developing, NICs, BRICs, MEDCs, LEDCs – which just goes to show how complex and ever-changing the global situation is.

Analysing data: measures of central tendency

When data are collected, they can be analysed in a variety of ways. In the last section, you attempted to describe sets of data using your own impressions and words. One way to analyse data is to look at 'typical' values or where the 'middle' of the data is. The set of data is then represented by a single quantity. This quantity (or statistical measure) is called a *measure of central tendency*. There are three measures of central tendency that are commonly used – the mean, the median and the mode – which you will investigate in this section.

Mode

The *mode* tells us about the values we are likely to find when we sample a set of data.

Investigation 1 – Determining the mode

criterion B

1 The *mode* is indicated in each of the following sets of data. Study each data set and figure out how to determine its mode.

Data	Mode
3, 5, 7, 8, 8, 10, 11, 12, 14	8
10, 10, 10, 14, 17, 18, 24, 24, 26, 31, 35	10
0 | 4 5 8 8 9 1 | 0 1 2 2 2 4 5 5 7 2 | 0 2 3 3 4 8 Key: 1 | 7 = 17	12
6.2, 8.5, 9.3, 6.2, 8.7, 11.4, 9.3, 7.6, 8.5	6.2 and 8.5
10 | 0 1 2 2 3 12 | 1 2 3 3 4 4 5 5 7 13 | 2 2 2 3 4 8 9 Key: 10 | 0 = 100	132
3, 15, 2, 7, 18, 10, 10, 7, 15	7, 10 and 15
12, 10, 17, 24, 31, 32, 18, 19, 25, 29	No mode

2 Write down a rule to determine the mode of a set of data.

3 Define *mode* as used in statistics.

Reflect and discuss 3

- Why would the mode be a useful statistic to represent a set of data? Explain.

- Is the mode always a value in the data set? Explain.

- Some would say that, if a data set has no values that repeat, every value is the mode. Does this make sense? Explain.

Median

Finding the median represents a different way of looking at a set of data values than finding the mode. You will investigate how to determine the median below.

Investigation 2 – Determining the median

1 The *median* is indicated in each of the following sets of data. Study each data set and figure out how to determine its median.

Data	Median
3, 5, 7, 8, 8, 10, 11, 12, 14	8
10, 10, 10, 14, 17, 18, 24, 24, 26, 31, 35	18
12, 9, 5	9
3, 5, 2, 7, 18, 10, 14, 21, 15	10
6.2, 8.5, 9.3, 6.2, 8.7, 11.4, 9.3, 7.6, 8.5	8.5
10 | 0 1 2 2 3 12 | 1 2 3 3 4 4 5 5 7 13 | 2 2 2 3 4 8 9 Key: 13 | 2 = 132	124
0 | 4 5 8 9 1 | 0 1 2 2 5 7 2 | 0 2 3 4 8 Key: 1 | 1 = 11	12

2 Write down a rule for finding the median of a set of data.

▶ Continued on next page

3 The *median* is again indicated in each of the following sets of data. Does your rule work for these data sets? If not, amend your rule to reflect any patterns that you see.

Data	Median
12, 9, 5, 2	7
3, 5, 7, 8, 8, 10, 11, 12, 14, 20	9
10, 10, 10, 14, 17, 18, 19, 24, 26, 31, 35, 42	18.5
3, 5, 2, 7, 18, 10, 14, 21, 15, 12	11
6.2, 8.5, 9.3, 6.2, 8.7, 11.4, 9.3	8.6
10 | 0 1 2 2 3 12 | 1 2 3 3 4 6 6 7 13 | 2 2 2 3 4 8 9 Key: 12 | 3 = 1230	1250
0 | 1 8 9 1 | 2 2 6 9 2 | 3 4 8 Key: 2 | 3 = 2.3	1.4

4 Write down a rule for finding the median of a set of data.

5 Define *median* as used in statistics.

Reflect and discuss 4

- Why would the median be a useful statistic to represent a set of data? Explain.

- Can there be more than one median? Explain.

- Is the median always a data value in the set? Explain using an example.

Example 2

Q Obesity rates can be an indication of the health of a country's population. The following data represent the percentage of adult men who are overweight in developed countries in the European Union.

▶ Continued on next page

Percentage of overweight men	Tally	Frequency
62%	||	2
63%	|||	3
64%	|	1
65%	|||	3
66%	|	1
67%	ⅢⅠ |	6

Find the median and mode of this set of data.

A The mode is 67%.

> The data values are already arranged in numerical order.

> The mode is the most common value in the list, which is the one with the highest frequency.

The median is 65%.

> The median is the middle value. Because there are 16 values, the median is the value in the middle of the 8th and 9th values. Starting with the lowest value of 62%, the 8th and 9th values are both in the category of 65%.

Practice 2

1 Find the median and mode of the following sets of data. Round answers to the nearest tenth where necessary.

a 24, 19, 21, 16

b 4.5, 2.2, 7.9, 4.5, 3.9, 8.4, 6.1

c 2, 7, 9, 5, 4, 2, 7, 4

d 67, 72, 80, 92, 80, 73, 69, 70, 84, 66

e 20 | 0 4 7 7 9
   21 | 1 1 1 5 6 6 7
   22 | 0 2 4 5 6 8 9

Key: 22|4 = 22 400

▶ Continued on next page

f

Number of hours of work	Tally	Frequency
4	‖‖	4
5	ℕ	5
6	ℕ ‖	7
7	‖‖	3
8	‖	2

g

```
 8 | 1  3
 9 | 4  7  8  8
10 | 0  1  1
11 | 5
```

Key: 8|3 = 8.3

h

Age (years)	Tally	Frequency
72	ℕ ℕ ℕ ℕ ℕ ℕ ‖	32
73	ℕ ℕ ℕ ℕ ‖‖	24
74	ℕ ℕ ℕ ‖	17

2 Data in the following table represent the percentage of the population in lower income countries in Latin America and the Caribbean that have access to electricity.

Country	Percentage of population with access to electricity (%)
Argentina	100
Belize	91
Brazil	100
Colombia	97
Costa Rica	99
Dominican Republic	98
Ecuador	97
El Salvador	94
Grenada	90
Guatemala	87
Haiti	38
Honduras	84
Jamaica	96
Mexico	99
Nicaragua	79
Paraguay	98
Peru	91
St Lucia	97
St Vincent and the Grenadines	96
Venezuela	99

▶ Continued on next page

a Find the median and mode of this data set.

ATL1

b Explain what these values tell you about the access to electricity for residents of these countries.

c Is there a value that you find surprising in this data set? Explain.

d If the lowest value is removed from the data set, find the new median and mode.

e How does removing one value change the measures of central tendency for this data set? Explain.

ATL2

f What must life be like in a country where access to electricity is difficult or even impossible? Explain.

3 Data in the following table represent the percentage of the population in higher income countries in Latin America and the Caribbean that have access to electricity.

Country	Percentage of population with access to electricity (%)
Antigua and Barbuda	95
Aruba	94
The Bahamas	100
Barbados	100
Bermuda	100
Cayman Islands	100
Curaçao	100
Puerto Rico	100
St Kitts and Nevis	99
St Martin (French)	100
Trinidad and Tobago	100
Turks and Caicos	93
Uruguay	99

a Find the mean and mode of this data set.

ATL1

b Explain what these values tell you about the access to electricity for residents of these countries.

c Do the results in questions **2** and **3** point to unequal access to electricity for citizens of low and high income countries? Explain.

4 a Find the median and mode of this set of data:

20   13   21   17   20   15   16   22   17

b Add an additional value that changes the median, but not the mode. Justify your choice.

▶ Continued on next page

c Add an additional value to the original data set that changes the mode but not the median. Justify your choice.

d Is it possible to add one value to the original data set that will change both the median and the mode? Justify your answer.

5 Create a set of data with the following characteristics:

a mode = 8, median = 12, number of values = 7

b mode = 25, median = 25, number of values = 8

c mode = 92, median = 68, number of values = 6

Mean

The *mean*, or *average*, of a set of data is a statistic that involves a mathematical calculation, as you will see in the next investigation. The symbol given to the mean is $\bar{x}$ (pronounced '*x*-bar').

Investigation 3 – Determining the mean

1 The mean ($\bar{x}$) is indicated in each of the following sets of data. Study each data set and figure out how to calculate its mean.

Data	Mean ($\bar{x}$)
12, 20	16
10, 9, 5	8
3, 8, 7, 6	6
10, 10, 13	11
6, 2, 1, 4, 12	5
3, 5, 2, 7, 16, 10, 14, 21, 12	10
6.2, 8.5, 9.3, 6.2, 8.7, 11.4, 9.3, 7.6	8.4
10 | 0 1 5 12 | 1 2 3 6 13 | 2 2 8 Key: 12 | 1 = 121	120
0 | 4 5 8 9 1 | 0 1 2 4 6 7 2 | 0 2 3 6 8 Key: 1 | 0 = 10	15

▶ Continued on next page

272   7 Univariate data

2 Describe how to find the mean in words.

3 Write down a rule/formula for calculating the mean ($\bar{x}$) or average of a set of data.

Reflect and discuss 5

- Why is the mean a useful statistic to represent a set of data? Explain.

- How can the mean be used in the calculation of the median? Explain with an example.

- The mean of a set of data is 12. Another piece of data is collected whose value is 15. How does this affect the mean? Explain.

Activity 2 – Summarizing your findings

1 While the mean, median and mode are all measures of central tendency, they have their similarities and differences. For the second and third columns, place a check (✓) in the column if it applies to the measure of central tendency.

2 Write down how to calculate each statistic in the fourth column.

3 Research online or in newspapers for an example in which each statistic is used. Write down your example (including the source of the example) in the last column.

Measure of central tendency	To make it easier to calculate, data must be written in numerical order	Statistic is always a value in the original data set	How to calculate	Example where it is used in real life (with source)
Mean ($\bar{x}$)				
Median				
Mode				

Practice 3

1 Find the mean ($\bar{x}$), the median and the mode of each set of data.
Round answers to the nearest tenth where necessary.

a 24, 31, 42, 37, 36, 24, 35, 40

b 7.6, 6.9, 7.8, 5.8, 6.2, 7.1, 6.9, 7.1, 6.6

c 100, 110, 108, 119, 117, 116, 121, 123

d 6, 9, 16, 14, 9, 5, 8, 7, 11, 13, 17, 4

e

20	2	4	6		
21	1	1	3	3	
22	1	2	7	9	9
23	4	5			

Key: 23|4 = 2340

f

16	1	3	9	9		
17	2	2	2	3	7	8
18	0	1	4	8	9	

Key: 17|2 = 17.2

g

Compulsory school attendance age	Frequency (number of countries)
12	2
13	3
14	3
15	8
16	11

2 Access to healthcare is an important part of living in any country.
The number of physicians per 1000 citizens in lower income countries
in Eastern Europe is given on the next page.

▶ Continued on next page

Country	Number of physicians per 1000 citizens
Albania	1
Armenia	3
Belarus	4
Bosnia and Herzegovina	2
Bulgaria	4
Kyrgyz Republic	2
Romania	2
Ukraine	4
Uzbekistan	3

a Find the mean ($\bar{x}$), median and mode of this data set. Round to the nearest tenth where necessary.

b Explain what these values tell you about the access to physicians for residents of these countries.

c Do these values seem high or low to you? Explain.

d Which measure(s) of central tendency best represent(s) the data? Justify your answer.

3 The number of physicians per 1000 citizens in higher income countries in Eastern Europe is given below.

Country	Number of physicians per 1000 citizens
Czech Republic	4
Estonia	3
Hungary	3
Latvia	4
Lithuania	4
Poland	2
Slovak Republic	3
Slovenia	3

a Find the mean ($\bar{x}$), median and mode of this data set. Round to the nearest tenth where necessary.

b Explain what these values tell you about the access to physicians for residents of these countries.

▶ Continued on next page

c Which measure(s) of central tendency best represent(s) the data? Justify your answer.

d Based on your answers to **2** and **3**, which measure(s) of central tendency could be used to show that there is *not* equal access to physicians in high and low income countries in this part of the world? Explain.

e Based on the your answers to **2** and **3**, which measure(s) of central tendency could be used to show that there *is* equal access to physicians in high and low income countries in this part of the world? Explain.

 f How would your family's life be different if you lived in a region where there was only 1 doctor per 1000 people? Explain.

4 Create a set of data with seven values where $\bar{x}$, the median and the mode are all different.

5 Create a set of data with at least eight values where the mean and median are the same, but the mode is different.

6 Create a set of data with eight values where the median and mode are the same, but the mean is different.

The effect of outliers

Sometimes, one data value seems to be very far away from the rest of the set of data points. Such a point is called an *outlier*. Outliers can occur when there is error in the collection of data, but they can also represent individuals who simply have extreme values. They are usually easy to spot, as in the following two graphs:

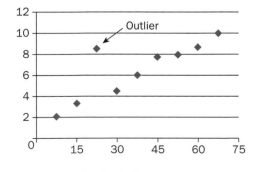

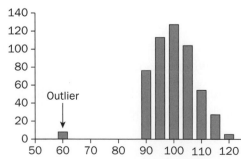

Activity 3 – The effect of outliers

For each of the following sets of data, determine which value is the outlier. Calculate the three measures of central tendency with all of the values in the set and then recalculate them with the outlier removed. Summarize your results in the table below.

Data set	Outlier	Mean of data set ($\bar{x}$)	Median of data set	Mode of data set	Mean with outlier removed ($\bar{x}$)	Median with outlier removed	Mode with outlier removed
2, 3, 4, 5, 5, 7, 9, 71							
8, 12, 14, 12, 9, 95, 10, 13							
24, 27, 26, 3, 27, 24, 20, 23, 25							
58, 15, 51, 60, 58, 58, 55							
101, 111, 107, 120, 110, 104, 118, 112, 26, 116, 121							

2 How does an outlier affect each of the three measures of central tendency? Summarize your findings in a table.

3 Which measure of central tendency is the *least* affected by an outlier? Which is the *most* affected? Explain why.

4 Explain how you think an outlier should be handled in a set of data when trying to analyse it.

5 Compare and contrast the three measures of central tendency. What are their strengths and weaknesses?

6 Under which circumstances would you choose the mean as the most representative measure of central tendency? Give an example to demonstrate what you mean.

7 Under which circumstances would you choose the median as the most representative measure of central tendency? Give an example to demonstrate what you mean.

8 Under which circumstances would you choose the mode as the most representative measure of central tendency? Give an example to demonstrate what you mean.

Formative assessment 1

Education is a fundamental human right and every child has the right to a quality education. However, not every country is able to fulfill this promise. In this task, you will compare low and high income countries from a region selected by your teacher to see if there is a difference in secondary school completion rates of adolescents.

criterion
C

▶ Continued on next page

1 Go to the data.unicef.org website and search for the data on 'secondary education'.

2 Download the data on lower secondary completion rates.

3 Write down the names of countries in the chosen region that are high-income and low-income.

4 Represent the lower secondary completion rate of these countries in two tables, one for low-income countries and one for high-income countries.

Analysis

1 Indicate if there are any outliers in your two data sets and explain what, if anything, you will do with them before calculating the measures of central tendency. Justify your decisions.

ATL1 2 Find the mean, median and mode of the data from high-income countries. Show your working.

ATL1 3 Find the mean, median and mode of the data from low-income countries. Show your working.

4 Which measure of central tendency is the most appropriate to use to represent the data sets? Justify your answer.

5 Do students in high and low income countries from the chosen region appear to have equal ability to complete lower secondary education? Explain.

ATL2 6 What would your life be like if you didn't complete lower secondary education (middle school)? Explain.

Action

1 Conduct research to find a program that aims to improve education levels in one of the countries in your study that had poor secondary school completion.

Analysing data: measures of dispersion

Whereas measures of central tendency describe the middle of a set of data values, *measures of dispersion* indicate how spread out the data are.

Range

The word 'range' often refers to something that is spread out.

Investigation 4 – Determining the range

1 The range is indicated in each of the following sets of data. Study each data set and figure out how to calculate its range.

Data	Range
12, 20	8
10, 9, 5	5
3, 8, 7, 6	5
10, 10, 13	3
6, 2, 1, 4, 12	11
3, 5, 2, 7, 16, 10, 14, 21, 12	19
6.2, 8.5, 9.3, 6.2, 8.7, 11.4, 9.3, 7.6	5.2
10 │ 0 1 5 12 │ 1 2 3 6 13 │ 2 2 8 Key: 13│2 = 132	38
0 │ 4 5 8 9 1 │ 0 1 2 4 6 7 2 │ 0 2 3 6 8 Key: 1│0 = 1.0	2.4

2 Describe how to find the range in words.

3 Write down a rule/formula for calculating the range of a set of data.

4 Verify your rule using two other data sets.

Reflect and discuss 6

- Why is the range a useful statistic to represent a set of data? Explain.

- Is the range necessarily one of the data values? Explain.

- What is the effect of an outlier on the range of a set of data? Explain with an example.

- Is it preferable for the range to be large or small? Explain.

Quartiles

The median divides a set of data in two halves. *Quartiles* divide a set of data into quarters: four sections with an equal number of values in each. There are three quartiles: the lower quartile (Q1), the middle quartile (the median, Q2) and the upper quartile (Q3).

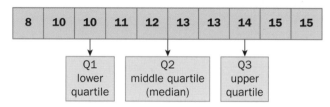

| 8 | 10 | 10 | 11 | 12 | 13 | 13 | 14 | 15 | 15 |

Q1
lower
quartile

Q2
middle quartile
(median)

Q3
upper
quartile

The difference between Q1 and Q3 is called the *interquartile range*. Data can be analysed using a *five-number summary*, which includes the minimum value, maximum value and the three quartiles.

Example 3

Q The unemployment rate of several developed countries, as a percentage of the total workforce, is given by the data below. This list includes the lowest unemployment rate as well as the highest unemployment rate of all the developed countries in the world, rounded to the nearest whole number.

0   12   6   8   26   13   11   7   8
6   9   15   4   7   10   3   6   9

Find the quartiles and interquartile range for this set of data.

A   0   3   4   6   6   6   6   7   7   8   8   9   9   10   11   12   13   26

Arrange the data in order, from lowest to highest.

0   3   4   6   6   6   6   7   7   8   8   9   9   10   11   12   13   26

median (Q2) = 7.5

Find the median. Since there is an even number of data values, the median is the average of the 9th and 10th value.

The median divides the 18 values in two groups of 9.

▶ Continued on next page

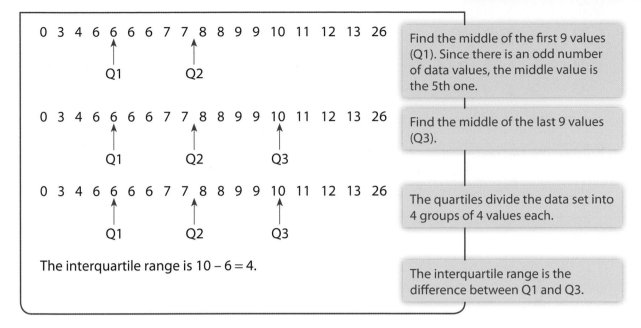

The interquartile range is 10 − 6 = 4.

| Find the middle of the first 9 values (Q1). Since there is an odd number of data values, the middle value is the 5th one. |

Find the middle of the last 9 values (Q3).

The quartiles divide the data set into 4 groups of 4 values each.

The interquartile range is the difference between Q1 and Q3.

Reflect and discuss 7

- What does a small or low interquartile range indicate? Explain.
- What percentage of values lie between Q1 and Q3? Explain.
- What percentage of values are above or below the median? Explain.

Practice 4

1 Find the range, five-number summary and interquartile range of each set of data. Round answers to the nearest tenth where necessary.

a 31, 28, 40, 25, 33, 42, 40, 38

b 15.2, 11.4, 13.9, 16.0, 12.8, 17.7, 20.4, 18.2, 16.5

c 310, 340, 360, 360, 300, 350, 320

d 2.8, 4, 3.5, 3.7, 2, 4.1, 5.6, 1.7, 1.1, 3.4, 2.8, 2.9

e

12	7 7 8
13	9
14	0 3 3 4
15	1 2 4

Key: 13|9 = 1390

▶ Continued on next page

f

```
43 | 0  0  2  3  4
44 | 1  2  5  6
45 | 2  4  4
```

Key: 43|2 = 432

g

Income share of lowest 20% of population (%)	Frequency (number of countries)
3	2
4	6
5	7
6	2
7	1
8	6

2 Arable land is land that is used for growing crops. The amount of arable land can indicate how able a country is to produce food to support its citizens. Countries with the highest percentage of arable land (as a percentage of total land area) are listed in the table below.

Higher-income countries		Lower-income countries	
Denmark	58%	Bangladesh	59%
Hungary	49%	Ukraine	56%
Czech Republic	41%	Moldova	55%
Isle of Man	39%	India	53%
Lithuania	38%	Togo	49%
Poland	36%	Burundi	47%
Germany	34%	Rwanda	47%
France	34%	The Gambia	44%

a Find the five-number summaries for the high and low income countries. Round to the nearest tenth where necessary.

b Explain what these values tell you about the availability of arable land in higher and lower income countries. Which category of country appears to have better availability of arable land?

▶ Continued on next page

c Is knowing the percentage of land that is arable sufficient information to determine a country's ability to grow its own crops? Explain. If not, what other factors would be important to research?

3 The number of people per public wi-fi hotspot, in those countries where hotspots are available, is summarized below. These represent the highest values among high-income countries.

Country	Number of people per public wi-fi hotspot
Canada	14
Singapore	320
Italy	979
Spain	13
Norway	12
Germany	10
Australia	11
United Arab Emirates	40
Poland	11
Japan	10
United Kingdom	6
Sweden	9

ATL1

a Find the range, quartiles and interquartile range of this data set. Round to the nearest tenth where necessary.

b Are there any outliers in this data set? Explain.

c Find the range, quartiles and interquartile range of the data set with any outliers removed.

d Which is most affected by the presence of outliers, the range or the interquartile range? Explain.

4 Create a set of data with an odd number of values (at least seven) where the three quartiles are all whole numbers.

5 Create a set of data with an even number of values (at least eight) where Q2 is *not* a value in the data set, but Q1 and Q3 are.

6 Create a set of data with at least six values where Q1 is greater than the range.

Representing data: box-and-whisker plots

Another form that can be used to represent data that includes some of the statistics you have calculated is called a *box-and-whisker plot* (sometimes referred to as a *boxplot*). While not all of the data values are clearly visible, a box-and-whisker plot allows you to see the range of the data and where each quartile is.

Example 4

Q The unemployment rate of several low-income countries, as a percentage of the total workforce, is given below. This list includes the lowest unemployment rate as well as the highest unemployment rate of all the low-income countries in the world, rounded to the nearest whole number.

0 30 30 5 28 13 11 41 42 23 95 46 31

A 0 5 11 13 23 28 30 30 31 41 42 46 95

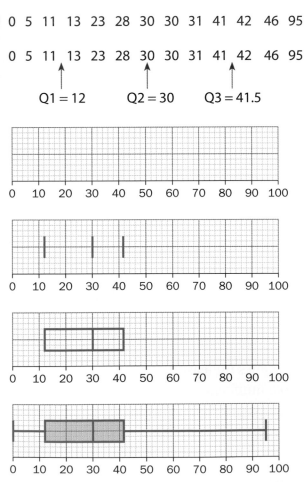

0 5 11 13 23 28 30 30 31 41 42 46 95

Q1 = 12   Q2 = 30   Q3 = 41.5

| Arrange the data in numerical order. |

| Find the quartiles. |

| Set up a number line with an appropriate scale to represent the data. |

| Use vertical lines above the number line to indicate the position of the quartiles. |

| Draw a box with Q1 and Q3 as its vertical sides. |

| Add in the 'whisker' by drawing a line from Q1 to the lowest value (0) and a line from Q3 to the highest value (95). |

| This box-and-whisker plot represents the data and demonstrates where 50% of the data points lie (between the quartiles/ in the box, also below the median and above the median). |

A box-and-whisker plot is a graphic representation of the
five-number summary, which includes the minimum value,
the maximum value and the three quartiles.

Reflect and discuss 8

- How is the range of a data set represented in a boxplot? Explain.
  The last two examples looked at the unemployment rates in
  low-income and high-income countries. Describe how their
  box-and-whisker plots would differ.

- A box-and-whisker plot of the percentage grades of two math
  classes achieved in a test is shown below. Which math class would
  you rather be in? Explain.

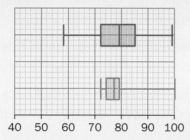

- Examples 3 and 4 looked at the unemployment rates in high-income
  and low-income countries. Redraw these box plots on the same
  set of axes. Compare these two box plots. If you were to write a
  headline summarizing this information, what would it be?

Practice 5

1 Find the range, the three quartiles and the interquartile range for
  data represented by the following box-and-whisker plots.

a

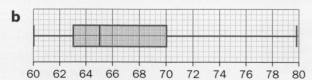

b

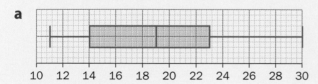

c

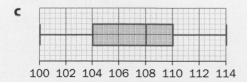

▶ Continued on next page

2 The following is a boxplot illustrating the distance (in kilometers) traveled by foot by children in a town in Nigeria to get to primary school.

a What is the shortest distance a child walked to school?

b What is the greatest distance a child walked to school?

c A distance of 1 kilometer by foot is considered to be the maximum a child should walk to school between the ages of 6 and 17. What percentage of children are walking 1 kilometer or less to school?

d Given this data set, what is the likelihood that the children in this town in Nigeria will continue to attend school on a regular basis? Explain.

ATL2

e What must life be like for someone who cannot go to school because it is too far away? What options does he/she have? What are the effects of those options? Explain.

3 Create a box-and-whisker plot to represent each of the following data sets. Show all of your working.

a 23, 28, 31, 20, 17, 29, 24

b 7, 10, 9, 13, 15, 8, 11, 10, 12

c 42, 50, 55, 43, 47, 47, 47, 51, 49, 44

d 3.2, 5.6, 7.1, 6.8, 5.0, 4.8, 5.1, 6.4

e

0	1 5 7 8
1	0 3 3 6 7 9
2	1 1 9
3	7 8 8

Key: 1|3 = 130

f

Percentage of seats in government held by women (%)	Frequency (number of countries)
42	4
41	2
40	2
39	4
37	6

Remember, if the data value 42 has a frequency of 4, that means some of the data are 42, 42, 42, 42.

▶ Continued on next page

4 The 20 highest birth rates (per 1000 people) of high-income and low-income countries are listed below.

Low-income countries	
Birth rate	Frequency
49	1
45	2
43	4
42	2
40	2
39	4
38	1
37	3
36	1

High-income countries	
Birth rate	Frequency
21	1
17	1
16	1
15	3
14	4
13	4
12	6

a Represent each set of data using a box-and-whisker plot. Draw both plots on the same axis.

b How does the representation show a difference in birth rates in high and low-income countries?

c What other information would you want to know in order to analyse the difference in birth rates between high and low-income countries?

5 The tables below list the ten higher-income and lower-income countries with the highest number of threatened species of birds.

Lower-income countries	
Country	Number of threatened species of birds
Brazil	165
Indonesia	131
Peru	120
Colombia	119
Ecuador	98
China	89
Philippines	89
India	84
Mexico	61
Bolivia	55

Higher-income countries	
Country	Number of threatened species of birds
United States	77
New Zealand	67
Australia	50
Japan	42
Chile	33
French Polynesia	32
Brunei Darussalam	24
Uruguay	22
Saudi Arabia	18
Singapore	17

▶ Continued on next page

a Represent each data set using box-and-whisker plots on the same set of axes. Show all of your working.

b Does there seem to be a difference between the number of threatened species of birds in higher and lower-income countries? Explain how your box-and-whisker plot represents your conclusion.

c Does being a lower-income country 'cause' bird species to be threatened? Explain.

d Propose two factors that could contribute to an increased number of threatened bird species.

6 a Create a box-and-whisker plot for a data set with a median of 20, a mean of 21 and a mode of 17.

b Create a new box-and-whisker plot for a data set with a median of 20, a mean of 21 and a mode of 20.

c How do changes in the values of the measures of central tendency affect a box-and-whisker plot? Explain.

Formative assessment 2

You have seen that there are some stark differences between countries in a number of aspects. In this task you will investigate whether there is a difference between these two groups in the percentage of the workforce that is female.

Conjecture

You will first make a conjecture (guess) about how you think the data will look.

1 Make a conjecture about what you think the range and spread of the data will be for both groups.

2 Explain why you chose the values you did.

▶ Continued on next page

Analysis

The data below represent the 12 lower-income countries with the highest percentage of the workforce that is female.

Country	Percentage of workforce that is female (%)
Rwanda	54.4
Mozambique	54.4
Burundi	51.7
Togo	51.2
Lao PDR	51.0
Nepal	50.8
Malawi	50.5
DR Congo	50.1
Ghana	50.0
Myanmar	49.7
Sierra Leone	49.5
Benin	49.4

1 Find the range, quartiles and the interquartile range for the data. Show all of your working.

2 Draw a box-and-whisker plot.

Research

1 Conduct research to find the 12 high-income countries with the highest percentage of the workforce that is female. Round values to the nearest tenth.

2 Find the range, quartiles and the interquartile range for the data. Show all of your working.

3 Draw a box-and-whisker plot for the data. Be sure to draw it on the same axes as the one for low-income countries.

4 Beneath the axes, write down the five-number summary for both sets of data.

Conclusion

1 Is there a difference in the percentage of the workforce that is female in high and low-income countries? Explain.

2 Which group seems to have a higher percentage of the workforce that is female? Why do you think that is?

3 Was your original conjecture correct? Explain.

Unit summary

A *stem-and-leaf* plot is a way to represent data that still preserves the original values. The stem represents one part of the number and the leaves another part. Stem-and-leaf plots should be accompanied by a key so that they can be interpreted correctly.

After exercise

8	1 3
9	4 7 8 8
10	0 1 1
11	5

Key: $8|3 = 8.3$

Before exercise

6	0 1 2 2 3 5 6 6
7	0 1 1 2 2 4 4 5 7 7 8
8	0

	9	
4 2	10	
8 4 1 0	11	
7 6 5 5 2 0 0	12	
7 3 1 1 0 0	13	
2	14	

Key: $10|2 = 102$

Measures of central tendency describe the middle of a set of data. There are three of them:

- The *median* is the middle value after the data are arranged in numerical order. If there is an even number of values, the median is the average of the two middle values.

- The *mode* is the most common value. There may be more than one mode, and there may be no mode if none of the values are repeated.

- The *mean* (or average) value is calculated by finding the sum of all of the data and dividing by the total number of values. The symbol for the mean is $\bar{x}$ (pronounced '*x*-bar').

Measures of dispersion describe the spread of a set of data:

- The *range* is the difference between the largest value and the smallest value.

- There are three *quartiles* that divide the data set into four sections with an equal number of values in each. The middle quartile (Q2) is the median, and is the middle of the data set (as seen above).

- The lower quartile (Q1) is the middle of the values less than the median. The upper quartile (Q3) is the middle of the values greater than the median.

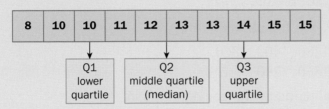

The *interquartile range* is the difference between Q3 and Q1 (Q3 − Q1).

The *five-number summary* includes the minimum value, the maximum value and the three quartiles.

A *box-and-whisker plot* is a way to represent the five-number summary of a set of data.

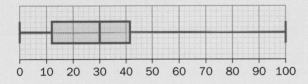

Unit review

criterion
A

> [] **Launch additional digital resources for this chapter**

Key to Unit review question levels:

| Level 1–2 | Level 3–4 | Level 5-6 | Level 7–8 |

1 **Draw** a stem-and-leaf plot (and key) for each of the given data sets.

  a 61, 72, 64, 70, 79, 83, 64, 61, 80, 65, 73, 79

  b 3.4, 5.1, 4.7, 5.5, 3.0, 3.9, 4.1, 5.5, 4.6, 4.9, 6.8

  c 1210, 1350, 1270, 1390, 1280, 1400, 1360,
  1260, 1350, 1200, 1410, 1560, 1200

2 **Find** the measures of central tendency for the data sets in question **1**. **Show** your working and give answers to the nearest tenth where necessary.

3 **Find** the range, quartiles and interquartile range for the data sets in question **1**.

4 The low-income countries with the highest level of spending on education (as a percentage of overall government spending) are represented in the stem-and-leaf plot below.

18	0 6 6 9
19	0 7
20	1 5 7
21	1 6
22	3
23	4
24	1
25	5

Key: $21\,|\,1 = 21.1$

  a What is the highest percentage of spending on education? What is the lowest?

  b Finland is one of the top developed countries in terms of spending on education, spending roughly 19% of its total government budget. How many countries spent more than Finland?

  c **Find** the mean, median and mode of the data set. Show your working and give answers to the nearest tenth where necessary.

5 Literacy rates can be a measure of the quality of the educational system in a country. Data for the 15 high-income countries and 15 low-income countries with the lowest literacy rates are summarized below.

High-income countries (%)	Low income countries (%)
95	19
95	27
97	30
97	36
98	37
98	38
98	38
98	39
99	40
99	43
99	44
100	48
100	48
100	49
100	52

ATL1 **a** Represent each set of data using a box-and-whisker plot. Draw both plots on the same axis.

ATL1 **b** Can you tell whether or not there is an unequal access to education in low and high-income countries? Which group seems to have better access? **Use** your representations to support your conclusion.

6 Students in high and low-income countries attend school for a specified number of days per year, as represented below.

Search for The Box Plot on **learner.org** for an interactive illustration of how a bar graph converts to a box plot.

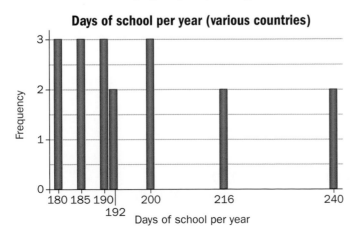

Days of school per year (various countries)

Days of school per year

a Convert the data to a stem-and-leaf plot as well as a box-and-whisker plot.

b What are the advantages and disadvantages of each of the three representations? **Explain**.

c Two of the countries in the bar graph are low-income countries. How long do you think their school year is? **Explain**.

d What other information would be important to know when comparing how long students in different countries are in school? **Explain**.

7 The population density is a measure of how many people live in a given area. The highest population densities in the world are given here, measured in the number of people per square kilometer.

Lower-income countries	
Maldives	1392
Bangladesh	1252
West Bank and Gaza	756
Mauritius	622
Lebanon	587
Republic of Korea (North Korea)	526
Rwanda	483
India	445
Comoros	428

Higher-income countries	
Monaco	19 250
Singapore	7909
Gibraltar	3441
Bahrain	1848
Malta	1365
Bermuda	1307
St Maarten (Dutch)	1177
Channel Islands	866
Barbados	663

a Are there any outliers in the two data sets? **Explain**.

b **Find** the measures of central tendency with and without the outlier(s).

c Which measure of central tendency do you think best represents the data? **Explain**.

d Do your calculations indicate an inequality between high and low-income countries when it comes to population density? **Explain**, being sure to state which group of countries benefits the most.

e Propose a reason for the differences between high and low-income countries.

f What must it be like to live in a place with **(i)** a high population density, **(ii)** a low population density? Which would you prefer? **Explain**.

8. The following is a box-and-whisker plot of the average annual income (Australian dollars, AUD) per person per country in the Oceania region.

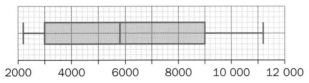

2000   4000   6000   8000   10 000   12 000

ATL2

a What is the minimum average annual income? What must it be like to live on that amount of money per year? **Explain**.

b 75% of all average annual incomes are below what amount?

c Two outliers were not included in the graph: one country with an average income of 50 000 AUD and the other with an average income of 58 000 AUD. Research the countries in the Oceania region. Which two countries do you think these outliers represent?

d If you created a new box-and-whisker plot to include the two countries in part **c**, **describe** how the new box-and-whisker plot would differ from the current one.

e Do you think it is preferable to have these two countries listed as outliers? **Explain**.

9. Access to improved sanitation facilities can have a large impact on the health of individuals and communities. The following tables show countries with the lowest percentage of the urban population that have access to improved sanitation.

Lower-income countries		Higher-income countries	
South Sudan	16%	Ireland	89%
Madagascar	18%	Guam	90%
Democratic Republic of Congo	20%	Latvia	91%
Ghana	20%	Trinidad and Tobago	92%
Sierra Leone	23%	The Bahamas	92%
Togo	25%	Cayman Islands	96%
Ethiopia	27%	Barbados	96%
Liberia	28%	Uruguay	97%
Uganda	29%	Lithuania	97%

ATL1

a **Find** the five-number summary for each set of data.

ATL1

b Represent each data set using a box-and-whisker plot. Plot the two graphs on the same axis.

c What does the box-and-whisker plot indicate about the equality of access to improved sanitation in urban areas? **Explain**.

d **Find** the measures of central tendency for the two sets of data.

e Which measure of central tendency do you think most accurately represents the data sets? **Explain**.

f Does your chosen measure of central tendency reveal an unfairness in access to improved sanitation between high and low-income countries? **Explain**.

g What other data or information might affect your decision as to whether or not there is inequality between these two groups? **Explain**.

ATL2

h What must it be like to live in a place where the access to improved sanitation is low, when you know that it is much higher in many other regions? **Explain**.

10 a **Create** a set of data with at least eight values that has a mean of 10, a median of 11 and mode of 8.

b Add an outlier that will increase the mean to 15.

11 a While doing research on developed countries and the percentage of homelessness, Mesfin found the average of ten countries to be 5%. After he included data on another country, the average jumped to 9%. What was the new data value that he found?

b With a mean value of 9% for eleven countries, Mesfin finds a country that has 3% homelessness. **Find** the mean value for the twelve countries.

12 GDP per capita generally shows the relative performance of countries, taking into account the cost of living within that country. In the Middle East, GDP per capita data (per thousand) are listed in the tables below.

Country	GDP per capita (per 1000)
Bahrain	30
Iran	18
Iraq	17
Israel	35
Jordan	11
Kuwait	71
Lebanon	19

Country	GDP per capita (per 1000)
Oman	44
Qatar	130
Saudi Arabia	54
Syria	3
Turkey	21
UAE	68
Yemen	3

a For the above data, **calculate** the mean GDP per capita for the Middle East region.

b Do you think Qatar is an outlier in this data set? **Justify** why or why not.

c Take Qatar out of the data and recalculate the mean GDP per capita.

d **Comment** on the effect that an extreme value has on the mean.

e Do some research and try to **explain** why Qatar has such a high value in comparison to the surrounding countries.

f **Draw** a box-and-whisker for this data set. For the purpose of this question, we will use a statistical definition of an outlier, which is, a data point that is either:

  $3 \times$ the interquartile range or more *above* the third quartile,

  or

  $3 \times$ the interquartile range or more *below* the first quartile.

Use this rule to reassess whether or not Qatar would be considered an outlier.

Summative assessment

How can you make a difference?

The World Bank Group, the largest development bank in the world that offers loans to lower-income countries, established a set of goals for sustainable development in 2017 with the collaboration of the United Nations.

Their first six goals include:

- ending poverty in all its forms by 2030
- ending hunger, achieving food security and improved nutrition, and promoting sustainable agriculture
- insuring healthy lives and promoting well-being for everyone, at all ages
- insuring equitable, quality education and promoting lifelong learning opportunities for all
- achieving gender equality and empowering all women and girls
- insuring availability of clean water and sanitation for all.

WEB LINK

There are 17 goals in total. You can read about the rest of them in the *Atlas of Sustainable Development Goals 2017 : From World Development Indicators* document on the World Bank Group website.

Pairs

You will be working with a partner and doing a comparison of low-income versus high-income countries to further your understanding of the conditions in other countries. Present your statistics on an infographic, using Canva, Infogram or Venngage, that will be submitted to your teacher, along with a companion report explaining your reasoning.

Part 1: Collecting the data

Go to The World Factbook compiled by the CIA (and/or similar site suggested by your teacher) to gather all of your data from different countries.

Your first decision as a pair will be to determine which ten high-income and ten low-income countries you will use in your comparison.

Select one of the sustainable development goals listed above as your focus, so that all of the statistics you will look at will be related to that goal.

You will be finding *at least three* sets of different data to make your comparisons. Make sure that the data you select can be directly compared for all countries. For example, look for percentages or 'per capita' statistics, as opposed to aggregate (total) number statistics. Represent your data using a table to make your analysis easier.

Part 2a: Analysing the data – an overview

For your 20 countries, find the data for the human development index statistic.

What is the human development index? Why do you think this is a good indicator to look at first to compare countries?

Create two box-and-whisker plots on the same axes, comparing human development in the high-income and low-income countries. Comment on the median, range and greatest spread.

Summarize your findings: were these results what you expected?

Part 2b: Analysing the data – your selected millennium development goal

Select and research three categories/indicators to complete your country comparisons that relate to your chosen development goal.

Explain how/why the three categories/indicators relate to your chosen goal.

Find data for your three categories for both high and low-income countries.

Display the data for the three categories using back to back stem-and-leaf plots with high-income countries on one side and low-income countries on the other.

Display each data for each category using a box-and-whisker plot. Plot data for high and low-income countries for each category on the same set of axes.

Which representation is better for comparing high versus low-income countries? Justify your decision.

Part 3: Reflection

Write a conclusion summarizing your results using the statistics you analysed above.

- What is the current situation in the high-income countries?
- How does it compare with that of the lower-income countries?
- Were the results what you were expecting? Explain why or why not.
- Do you think the World Bank will be able to meet its goal?
- What role do you see the developed world playing in the success of the goal?
- How can we use information to instigate change and make a difference?

Part 4: Action – How can you make a difference?

There are many charities working to help communities in low-income countries. Research some of the organizations and charities that are working to achieve the development goal you have chosen.

With your partner, decide which organization you feel is doing a good job to address the needs of the low-income countries you analysed. Create a basic action plan outlining what you could do to help them by supporting a particular initiative or activity.

Part 5: Presenting your ideas

You will create a one or two page infographic outlining:

- your comparisons (including stem-and-leaf plots and box-and-whisker plots and measures of central tendency)
- conclusions
- action plan ideas

Part 6: Follow-through

Every group has 3 minutes to present their ideas to the class using the infographic as support. Once hearing every group's action plan and ideas, as a class you may want to get involved and support a particular initiative!

Answers

Unit 1

Practice 1

1 **a** Simplified   **b** $2:3$ **c** $3:4$ **d** $1:2$ **e** $20:1$ **f** $1:15$
   g Simplified   **h** Simplified   **i** $19:1$ **j** Simplified   **k** $503:494$ **l** $4:1$

2 Individual responses, example answers include:
   a $4:10, 6:15, 8:20$ **b** $1:19, 4:36, 6:54$ **c** $1:3, 2:6, 3:9$ **d** $2:5, 4:10, 6:15$
   e $12:14, 18:21, 24:28$ **f** $1:3, 2:6, 3:9$ **g** $8:1, 16:2, 24:3$ **h** $4:1, 8:2, 16:4$

3 **a** Individual response, for example:
   DC: 14, Marvel: 10
   DC: 21, Marvel: 15
   b No   **c** 65

4 **a** Individual response, for example a picture of 2 green flowers, 3 blue flowers and 5 red flowers
   b Individual response, for example a picture where there are twice as many girls as there are boys
   c Individual response, for example a picture where there are exactly as many trees as there are bushes

5 $4:5, 12:15, 28:35$ are equivalent
   $2:8, 1:4, 12:48$ are equivalent
   $1:3, 4:12, 7:21$ are equivalent
   $5:10, 4:8, 9:18$

6 Individual response

7 30 Pokémon cards, 12 trainer cards, 18 energy cards

8 **a**

Country	Women's wages: Men's wages for same work (currency)	Simplified Ratio
Bulgaria	$1.5:2$ (lev)	**3:4**
Australia	$0.83:1$ (dollar)	**83:100**
Russia	$2.40:3$ (ruble)	**4:5**
United Arab Emirates	$0.99:1$ (dirham)	**99:100**

   b Bulgaria   **c** Women most disadvantaged in Bulgaria; least disadvantaged in The United Arab Emirates

9 **a** 27, 63   **b** 11, 22, 55   **c** 18, 6, 8   **d** 12, 24, 30, 54

10 294 Swedish, 462 Danish, 273 Finnish

11 **a** $k = 8$ **b** $p = 40$ **c** $x = 21$ **d** $r = 21, s = 49$ **e** $q = 3.5$ **f** $z = 1.5$

Practice 2

1   $112:564, \dfrac{28}{141}, 19.9\%$

   $42:70, \dfrac{3}{5}, 60\%$

   $24:30, \dfrac{4}{5}, 80\%$

   $23:46, \dfrac{1}{2}, 50\%$

   $2.5:6, \dfrac{5}{12}, 41.7\%$

   $17.5:21, \dfrac{5}{6}, 83.3\%$

   $8:72, \dfrac{1}{9}, 11.1\%$

   $18:54, \dfrac{1}{3}, 33.3\%$

2 a Individual response

b Women: 12:13, 0.48, 48%
Men: 13:12, 0.52, 52%
African American: 7:43, 0.14, 14%
Hispanic or Latino/ a: 13:87, 0.13, 13%
Native American or Pacific Islander: 3:97, 0.03, 3%
White: 12:13, 0.48, 48%
Asian American: 11:39, 0.22, 22%
English Language Learners: 4:21, 0.16, 16%

c 48% of the students admitted to the university are female, whereas 52% of the students are male. These are approximately equal.

3 a USSR: $3:8, \dfrac{3}{8}, 0.375, 37.5\%$

Canada: $4:8, \dfrac{1}{2}, 0.5, 50\%$

b Individual response   **c** $32:31, \dfrac{32}{31}, 1.03, 103.2\%$

d Individual response

4 a Individual response   **b** 33.3%   **c** 67.2%   **d** $\dfrac{1}{16}, 16$

Practice 3

1 a $w = 8$ **b** $x = 21$ **c** $p = 15$

d $m = 9$ **e** $h = 33$ **f** $x = \dfrac{285}{8} = 35.63$

g $a = \dfrac{2600}{81} = 32.10$ **h** $b = \dfrac{279}{5} = 55.8$ **i** $y = \dfrac{70}{3} = 23.33$

j $w = \dfrac{96}{5} = 19.2$ **k** $v = \dfrac{147}{10} = 14.7$ **l** $z = \dfrac{7}{9} = 0.78$ **m** $c = \dfrac{289}{15} = 19.27$

n $f = \dfrac{84}{73} = 1.15$ **o** $k = \dfrac{28}{5} = 5.6$ **p** $y = \dfrac{288}{11} = 26.18$

2 a Equivalent   **b** Not equivalent   **c** Not equivalent   **d** Not equivalent

3 a 705 m   **b** $\dfrac{7}{3}$ cm   **c** 1185 cm

4 a 9 600 km   **b** 20 years   **c** 298   **d** 1542

5 21

Practice 4

1 a No   **b** Yes, $Y = 4X$ **c** No

d Yes, $Y = \dfrac{1}{2}X$ **e** No   **f** Yes, $Y = 10X$

2 The song will still take 4 minutes to sing

3 $4.67

4 a Yes   **b** 25000   **c** $200 000

5 a Constant of proportionality $\dfrac{25}{8}$ (people per metre of length of chain)

b Line passing through origin with gradient $\dfrac{25}{8}$

$$Y = \dfrac{25}{8}X$$

c 640 000 m

6 a Store 1: $1.87 per box, Store 2: $1.86 per box so Store 2 is the better buy

b Yes

c 15 boxes from Store 1 = $28.05; 15 boxes from Store 2 = $27.90

d Individual response

Unit Review

1 a $3:4$ b $5:1$ c $1:3$ d Already simplified   e $3:2$ f $4:3$

2 Individual response, example answers include:
   a $2:4$, $3:6$ b $8:18$, $12:27$ c $1:2$, $2:4$ d $2:12$, $3:18$ e $8:2$, $12:3$ f $2:3$, $4:6$

3 a Individual response, which suggests that mullets with the largest 'party length' to 'business length' are most likely to win
   b $6:2$

4 a College graduates: $\dfrac{9}{25}$ b $0.36:0.34:0.3$ c $18:17:15$
   High–school graduates: $\dfrac{17}{50}$
   Neither: $\dfrac{3}{10}$

5 a Yes   b 3   c 72   d 36

6 a Syria $1:5$
   South Sudan $1:2$
   Ethiopia $1:10$
   Democratic Republic of Congo: $1:5$
   b No, it would be possible to make a prediction but, as the relationship is not necessarily proportional, it would not be possible to accurately determine the winner.
   c Yusra

7 a $p = 45$ b $m = 18$ c $y = 3.7$ d $t = 6.6$ e $f = 1.5$ f $k = 21.9$

8 a Male: 1100, Female: 715   b $363:37$

9 $2.392 million ($2 392 000)

10 $5:6$

11 a $1900:11$, $\dfrac{190}{11}$, 17.27 b 4543   c Individual response (e.g. shortage of firefighters)
   $2300:263$, $\dfrac{2300}{263}$, 8.75

12 a Michael Johnson
   b Individual response (Based on average speed alone, Johnson is quicker)
   c Individual response (There was no race over the same distance from which a direct comparison could be made. A more sophisticated response may also indicate that the acceleration phase of a race is likely to bring the average speed over 100 m down more than the average speed over 200 m).
   d Johnson: 14.49s
   Bailey: 14.76s
   e This average speed slower than his average speed for the 100m race

13 a Individual response, for example:
   b Yes
   c $P = \dfrac{B}{978}$ where P is the number of picograms and B is the number of base pairs in millions
   d 11.5

Number of Picograms	Number of Base Pairs (millions)
1	978
2	1956
3	2934
4	3912

Unit 2

Practice 1

1 a Heads, Tails   b 1, 2, 3, 4, 5, 6
   c Unknowable – would depend on the number of classes/periods   d 1, 2,…, 27, 28 (and in a leap year, 29)

2 a List: Italian & Laser Tag, Italian & Go-Cart, Italian & Arcade, Japanese & Laser Tag, Japanese & Go-Cart, Japanese & Arcade, Moroccan & Laser Tag, Moroccan & Go-Cart, Moroccan & Arcade

b 9   **c** Individual response

d You would have an additional three items in the sample space; the activity can be paired once with each of the restaurants

e Italian & Laser Tag ($115), Italian & Go-Cart ($103), Italian & Arcade ($85), Japanese & Laser Tag ($128), Japanese & Go-Cart ($116), Japanese & Arcade ($98), Moroccan & Laser Tag ($122), Moroccan & Go-Cart ($110), Moroccan & Arcade ($92)

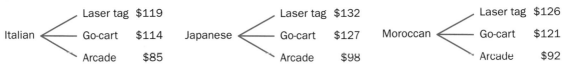

	The option includes go-carting	The option does not include go-carting
Celebration cost is under $120	Italian & Go-Cart	Italian & Laser Tag, Italian & Arcade, Japanese & Arcade, Moroccan & Arcade
Celebration cost is over $120	Japanese & Go-cart; Moroccan & Go-cart	Japanese & Laser Tag, Moroccan & Laser Tag

g 5   **h** 3   **i** 1

j The combination of Italian and go-carting is the least expensive, but go-carting with either of the other dinner options isn't much more expensive.

3 a 324

b Sample space is as before, except with the addition of 'Dr Orchid' in the suspects list and 'poison' in the weapons list 441

4 a Individual response   **b** 16 outcomes

c Possible scores are 0, 1, 2, 3, 4, 5, 6, 7, 8, 9 and 10
The scores of 0, 1, 2, 8, 9 and 10 each occur in only one way; the scores of 3, 4, 5, 6, and 7 can each occur in two ways.

Practice 2

1

Probability Statement	How likely is this event?	Make up a scenario. What decision might you make given this probability value?
P(rain) = 0.23	Unlikely	Individual response
P(red marble) = $\frac{3}{4}$	Likely	Individual response
P(perfect score) = 80%	Likely	Individual response
P(win) = 0.5	Equally likely as unlikely	Individual response
P(girl) = $\frac{3}{31}$	Unlikely	Individual response

2 a i 0.5, 50%, equally likely as unlikely   **ii** 0.25, 25%, unlikely
   iii 0.375, 37.5%, unlikely   **iv** 0.625, 62.5%, likely
    v 0.125, 12.5%, unlikely   **vi** 1, 100%, certain

b Individual response

3 a Likely   **b** Certain   **c** Unlikely   **d** Unlikely

4 Individual response

Practice 3

1 a

Bus — Walk, Cycle   Taxi — Walk, Cycle   **b** $\frac{1}{4}$

2 No, she is incorrect. Rolling an odd or even number are equally as likely. Both probabilities are $\frac{3}{6} = \frac{1}{2} = 0.5 = 50\%$

3 a $\dfrac{1}{3700000} = 0.00000027 = 0.000027\%$　　　**b** (Very) Unlikely

4 a i $\dfrac{1}{8}$　　　**ii** $\dfrac{7}{8}$

　　b The minimum number of coins required is 5　　**c** 11

5 a i $\dfrac{1}{37}$　　**ii** $\dfrac{18}{37}$　　**iii** $\dfrac{18}{37}$　　**iv** $\dfrac{6}{37}$　　**v** $\dfrac{1}{37}$

　　b i $\dfrac{22}{36} = \dfrac{11}{18} = 0.61$ (2 d.p.)

　　ii $\dfrac{22}{37}$

　　iii There are now 37 possible outcomes rather than 36
　　c 1655　　　　**d** Individual response

6 a $\dfrac{7}{12}$　　　**b** $\dfrac{1}{12}$　　　**c** $\dfrac{4}{11}$

7 a $\dfrac{1}{3}, \dfrac{2}{3}$　　　**b** $\dfrac{1}{2}$

　　c The optimal strategy is to change. The initial probability of having chosen the winning door is $\dfrac{1}{3}$. If the
　　contestant insists on sticking with their choice, this probability is unaffected by any further action. Suppose
　　instead the contestant insists on switching door. If the contestant initially chooses the winning door, they will
　　then switch to a losing door. However, if the contestant initially chooses either of the losing doors, they will
　　then switch to the winning door. Thus the switching strategy carries a greater probability of $\dfrac{2}{3}$.

Practice 4

1 a i $\dfrac{1}{24}$　　**ii** $\dfrac{1}{3}$　　**iii** $\dfrac{1}{12}$　　**iv** $\dfrac{1}{6}$

　　v $\dfrac{1}{6}$　　**vi** $\dfrac{5}{24}$

　　b Examples of complementary events:
　　P(dollar amount), P(no dollar amount)
　　P(Lose a turn), (P not Lose a turn)
　　P(ends in 50), P(does not end in 50)

　　c Individual response

2 a List: 2, 3, 4, 5, 6, 7, 8, 9, 10, 11, 12

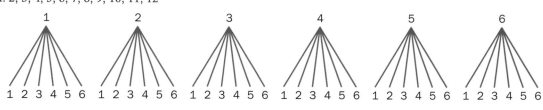

Total　　2 3 4 5 6 7　　3 4 5 6 7 8　　4 5 6 7 8 9　　5 6 7 8 9 10　　6 7 8 9 10 11　　7 8 9 10 11 12

	1	2	3	4	5	6
1	2	3	4	5	6	7
2	3	4	5	6	7	8
3	4	5	6	7	8	9
4	5	6	7	8	9	10
5	6	7	8	9	10	11
6	7	8	9	10	11	12

　　b 7
　　c i $\dfrac{1}{2}$　　**ii** $\dfrac{15}{36}$　　**iii** $\dfrac{5}{6}$　　**iv** $\dfrac{2}{9}$

d **i** An even number appearing next, $\frac{1}{2}$ **ii** A composite number appearing next, $\frac{7}{12}$

    iii 7 appearing next, $\frac{1}{6}$ **iv** Neither 7 nor 11 appearing next, $\frac{7}{9}$

3 **a** 7 is the most probable outcome, so selecting 7 as the losing sum makes the game more difficult to win

  b No

  c Either a 6 or an 8, since after 7 these are (equally) the most probable outcomes

  d Individual response, for example: no, 7 is the most likely outcome so the odds are against the player

4 **a** Individual response (e.g. H = Heads, T = Tails, List: HH, HT, TH, TT)

  b $\frac{1}{4}$ **c** $\frac{1}{2}$ **d** $\frac{1}{4}$

5 **a** P(odd) $= \frac{1}{2}$

     P(number > 40) $= \frac{5}{9}$

     P(multiple of 5) $= \frac{1}{5}$

     P(prime number) $= \frac{4}{15}$

     P(repeated digit) $= \frac{4}{45}$

     P(one of your numbers is called) $= \frac{3}{18}$

     P(one of your numbers is *not* called) $= \frac{15}{18}$

  b 'One of your numbers is called' and 'One of your numbers is *not* called'

     $\frac{3}{18} + \frac{15}{18} = 1$

  c Individual response

Practice 5

1 **a** 1, 2, 3, 4   **b** Monday, Tuesday, Wednesday, Thursday, Friday

  c HH, HT, TH, TT   (H = Heads, T = Tails)   **d** The six paths

2 **a** $\frac{1}{4}$ **b** $\frac{2}{5}$ **c** $\frac{1}{2}$ **d** $\frac{1}{2}$

3 **a** $\frac{1}{6}$ **b** $\frac{5}{6}$ **c** $\frac{1}{6} + \frac{5}{6} = 1$

4 **a** P(Win) $= \frac{1}{3}$

     P(Lose) $= \frac{2}{3}$

  b P(Win) $= \frac{1}{3}$

     P(Lose) $= \frac{2}{3}$

  c Individual response (students should make some reference to the theoretical and experimental probabilities being the same)

5 **a** Experimental probability is $\frac{1}{4}$

     Theoretical probability is $\frac{1}{3}$

     So the experimental probability is less than the theoretical probability

  b Theoretical probability is $\frac{1}{3}$

     Experimental probability is $\frac{2}{5} = 0.4$

  c You would expect the experimental probability to decrease, since the greater the number of trials, the closer you expect the experimental probability to be to the theoretical probability.

6

1 ⟨ Head / Tail   2 ⟨ Head / Tail   3 ⟨ Head / Tail   4 ⟨ Head / Tail

5 ⟨ Head / Tail   6 ⟨ Head / Tail   7 ⟨ Head / Tail   8 ⟨ Head / Tail

P(heads and odd) $= \dfrac{1}{4}$

7 Individual response

Practice 6

1 Individual response

2 $\dfrac{27}{64} = 0.422$

Individual response

3 Individual response

4 a 0, 8, 9, 10, 11, ..., 24   **b** $\dfrac{1}{6}$ **c** $\dfrac{5}{6}$

d $\dfrac{25}{36}$ **e** $\dfrac{125}{216}$ **f** $\dfrac{125}{1296}$

g Individual response (answers should include reference to the trade-off between the greater reward from continuing rolling and the associated probability of rolling a 1)

h $\dfrac{1}{216}$ **i** $\dfrac{1}{7776}$ **j** Yes

Unit Review

1 a Individual response (Tree diagram or table)
 b Individual response (List or tree diagram)
 c Individual response (List or table)

2 a $\dfrac{1}{4}$

 b

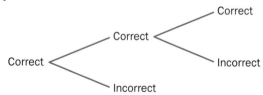

Incorrect

List: C = Correct, I = Incorrect
I, CI, CCI, CCC

 c $\dfrac{1}{2}$

3 a i $\dfrac{1}{8}$ **ii** $\dfrac{3}{8}$ **iii** $\dfrac{1}{2}$ **iv** $\dfrac{1}{2}$ **v** $\dfrac{1}{8}$

 b $\dfrac{5}{8} + \dfrac{3}{8} = 1$ **c** Individual response

 d (a) (i): unlikely, (ii): unlikely, (iii): equally likely as unlikely, (iv): equally likely as unlikely, (v): unlikely
  (b) certain

4 a i $\dfrac{1}{26}$ **ii** $\dfrac{4}{13}$ **iii** $\dfrac{8}{13}$

  iv $\dfrac{3}{26}$ **v** $\dfrac{8}{13}$ **vi** 0

b i $\frac{14}{15}$ **ii** $\frac{1}{15}$ **iii** $\frac{4}{15}$ **iv** $\frac{4}{15}$

c Individual response

5 a i $\frac{2}{3}$ **ii** 1   **iii** $\frac{1}{3}$ **iv** $\frac{1}{6}$

b Sample space:

Attacker rolls	Defender rolls	Battle 1 won by	Battle 2 won by	War won by
2, 5	1, 1	Attacker	Attacker	Attacker
	1, 2	Attacker	Attacker	Attacker
	1, 3	Attacker	Attacker	Attacker
	1, 4	Attacker	Attacker	Attacker
	1, 5	Defender	Attacker	Defender
	1, 6	Defender	Attacker	Defender
	2, 1	Attacker	Attacker	Attacker
	2, 2	Attacker	Defender	Defender
	2, 3	Attacker	Defender	Defender
	2, 4	Attacker	Defender	Defender
	2, 5	Defender	Defender	Defender
	2, 6	Defender	Defender	Defender
	3, 1	Attacker	Attacker	Attacker
	3, 2	Attacker	Defender	Defender
	3, 3	Attacker	Defender	Defender
	3, 4	Attacker	Defender	Defender
	3, 5	Defender	Defender	Defender
	3, 6	Defender	Defender	Defender
	4, 1	Attacker	Attacker	Attacker
	4, 2	Attacker	Defender	Defender
	4, 3	Attacker	Defender	Defender
	4, 4	Attacker	Defender	Defender
	4, 5	Defender	Defender	Defender
	4, 6	Defender	Defender	Defender
	5, 1	Defender	Attacker	Defender
	5, 2	Defender	Defender	Defender
	5, 3	Defender	Defender	Defender
	5, 4	Defender	Defender	Defender
	5, 5	Defender	Defender	Defender
	6, 1	Defender	Attacker	Defender
	6, 2	Defender	Defender	Defender
	6, 3	Defender	Defender	Defender
	6, 4	Defender	Defender	Defender
	6, 5	Defender	Defender	Defender
	6, 6	Defender	Defender	Defender

c $\frac{29}{36}$

d If the attacker rolls a 1 and any other number then the defender will either win both battles, or win one and lose one, and in the event of such a tie the defender wins.

6 Individual response

Unit 3

Practice 1

1. **a** No **b** Yes **c** No **d** Yes **e** Yes **f** No
   g Yes **h** Yes **i** Yes **j** No **k** No **l** Yes

2. **a** 12 **b** −7 **c** 82 **d** 3 **e** −26 **f** 103

3. **a** < **b** > **c** > **d** > **e** > **f** >
   g < **h** > **i** < **j** < **k** > **l** >

4. **a** Individual response (e.g. reference to a scale in which there is both 'negative brightness' and 'positive brightness'.)
   b Jupiter and Halley's Comet
   Individual response
   c Individual response

5. **a** + **b** − **c** + **d** − **e** + **f** +

6. **a** 9, −1, −3, −5, −7
   b 8, 4, 0, −2, −4, −12, −20
   c 4, 2, 0, −3, −5, −9, −10
   d −31, −34, −45, −49, −52, −67
   e −101, −104, −112, −119, −121, −125, −134

7. **a** Mount Everest, Cusco, Nairobi, Paris, Manila, Amsterdam, Baku, Deep Lake, Death Valley, Marianas Trench
   b Student's own sketch. Would be ideal if it was set vertically, given the context.
   c Individual response
   d Mount Everest is higher than the Marianas Trench is deep
   e Individual response
   f Individual response

Practice 2

1. **a** 2 **b** 12 **c** 8 **d** 7 **e** 34
   f 102 **g** 79 **h** 41 **i** 11 **j** 0

2. **a** > **b** > **c** < **d** = **e** > **f** <
   g > **h** < **i** < **j** < **k** < **l** =

3. **a** Fifth Avenue would represent 0
   b Same distance
   c Include negative integers to represent house numbers in West 41st streets
   d Individual response
   e Individual response

4. **a** −105 m assuming 0 is sea level
   b The absolute value is the distance to sea level
   c Individual response

5. **a** South Pole, Framheim, Whale Bay, Funchal, Kristiansand
   b Whale Bay, Framheim, Kristiansand, Funchal, South Pole
   c 22 degrees Celsius
   Individual response
   d Individual response (e.g. 12, −12, 13, −13)

Practice 3

1. $A(2, 4)$ first quadrant
   $B(0, 3)$ y-axis
   $C(−1, 2)$ second quadrant
   $D(2, −3)$ fourth quadrant
   $E(4, 1)$ first quadrant
   $F(0, 0)$ origin
   $G(−5, 0)$ x-axis
   $H(0, −2)$ y-axis

2

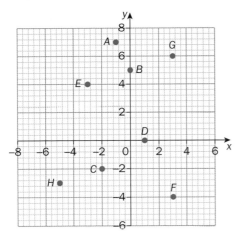

3 Individual response

Practice 4

1 **a** −32　**b** 21　**c** −60　**d** 45　**e** −44
　f 48　**g** 49　**h** −54　**i** 2　**j** 35
　k −121　**l** −27　**m** 36　**n** 48　**o** −48
　p 10　**q** −24　**r** 70　**s** 6　**t** −9

2 −12: −2 × 6, −12 × 1, −3 × 4, −4 × 3
　−20: 10 × (−2), −20 × 1, −4 × 5
　12: −3 × (−4), −1 × (−12), −6 × (−2)
　20: 2 × 10, −5 × (−4), −2 × (−10), −1 × (−20)

3 **a** 6　**b** −4　**c** −3　**d** 9　**e** −7　**f** −6

4 Individual response

5 Individual response

6 **a** Danakil Depression, Salton Sea, Deep Lake, Caspian Sea Depression, Baku, Laguna Salada, Lammefjord, Kristianstad
　b 28, the depth of Baku relative to sea level
　c **i** Five times　**ii** Twice　**iii** Four times　**iv** 12.5 times
　　　　　　　　　　　　　　　　　　　　　　　　　　　Not an integer

7 **a** −8, −6, −4, −3, −2, −1, 1, 2, 3, 4, 6, 8　**b** −9, −6, −4, −3, −2, −1, 1, 2, 3, 4, 6, 9
　c −10, −6, −5, −4, −3, −2, −1, 1, 2, 3, 4, 5, 6, 10　**e** −8, −4, −2, −1, 1, 2, 4, 8

8 **a** 66　**b** 84　**c** −54　**d** 21　**e** −8　**f** 7
　g −52　**h** −112　**i** −90　**j** −100　**k** −112　**l** 77
　m 32　**n** −12　**o** −10　**p** −24　**q** 36　**r** 420
　s −54　**t** −27　**u** −64　**v** 25　**w** −1　**x** 10 000

9 **a** Viking: −6 364 970; Cassini-Huygens: −9 430 500; Voyager 1: −6 329 700; Hubble: −2 592 330; Apollo: −29 700 000

10 181.67 minutes

Practice 5

1 **a** −3 + 6 = 3　**b** 5 − 4 = 1　**c** −7 − (−2) = −5　**d** −3 − (1) = −4

2 **a** 1　**b** −6　**c** −5　**d** 6　**e** −13　**f** 2
　g −8　**h** 6　**i** −9　**j** −39　**k** −13　**l** 9
　m 4　**n** −4　**o** −2　**p** −4　**q** −11　**r** −11

3 −7, −6, −4, 1, 2, 3, 4

4 **a** −1　**b** 2　**c** −6　**d** −10　**e** 2
　f 8　**g** 3　**h** −5　**i** 2　**j** 9

5 **a** 17 Degrees Celsius　**b** −18 Degrees Celsius

6 −31 Degrees Celsius

7 −3 m

8 19 842 m

9 a 14   **b** Individual response

10 550 metres below sea level

11 102 Degrees Celsius

12 a 3   **b** 4   **c** 1   **d** 5   **e** 2   **f** 6
   so f, d, b, a, e, c

Practice 6

1 a −11   **b** −39   **c** −14   **d** 70   **e** −4
   f −5   **g** 6   **h** −20   **i** −2   **j** 10
   k 6   **l** $\frac{1}{9}$ **m** 20   **n** 82   **o** −8
   p $-\frac{23}{21}$ **q** 63   **r** 3   **s** −54   **t** 28

2 88 kg

3 −29 Degrees Celsius

4 a $(-10-6) \div (2 \times (-4)) = 2$ **b** $9 + 2 - (6 \div (-3)) = 13$ **c** $(7 - 4 - (12 \div (-2))) \times 3 = 27$

Unit Review

1 a <   **b** >   **c** >   **d** <   **e** >   **f** <

2 a −   **b** −   **c** −   **d** +   **e** +   **f** +

3 a −7, −3, −1, 0, 2   **b** −12, −6, −2, 1, 5, 7, 9   **c** |0|, |−1|, |−2|, |−4|, |8|, |−10|

4 $A(5, 6)$; $B(−4, 1)$; $C(0, 6)$; $D(5, 0)$; $E(−6, 2)$; $F(−4, −3)$; $G(3, −1)$; $H(−2, 5)$; $I(2, 3)$; $J(−5, 6)$; $K(−1, −1)$; $L(4, −4)$

5

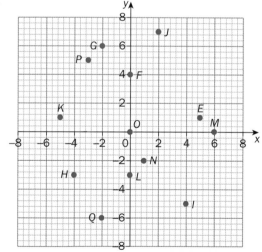

6 a Individual response   **b** Individual response

7 a

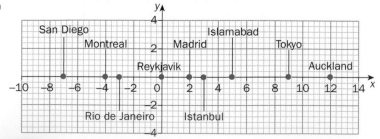

b Istanbul, Rio de Janerio

c Reykjavik, Madrid, Istanbul & Rio de Janeiro, Montreal, Islamabad, San Diego, Tokyo, Auckland

d Individual response

8 a 15   **b** −11   **c** −3   **d** −88   **e** −7

   f −2   **g** −4   **h** −33   **i** 24   **j** 9

   k 18   **l** −42   **m** −10   **n** $\frac{5}{8}$ **o** −6

   p −5   **q** −4   **r** −6   **s** −21   **t** −5

   u −6   **v** 30   **w** −33   **x** −4   **y** −16

9 (3, −5) lies in the fourth quadrant with x-coordinate 3 and y-coordinate −5 whereas (−5, 3) lies in the second quadrant with x-coordinate −5 and y-coordinate 3

10 a 40 Degrees Celsius   **b** King William Island by 9 Degrees Celsius

   c Christiana and King William Island by 48 Degrees Celsius   **d** Individual response

11 a −11   **b** −3   **c** 3   **d** 11   **e** 17   **f** −2

   g −28   **h** 5   **i** 3   **j** −6   **k** 5

12 a 10 Degrees Celsius   **b** −15 Degrees Celsius   **c** −20 Degrees Celsius

   d −25 Degrees Celsius   **e** −40 Degrees Celsius

13 Individual response

14 a Great Barrier Reef & Victoria Falls, Machu Picchu, Angkor Wat, Blue Hole, Chichen Itza, Petra, Stonehenge

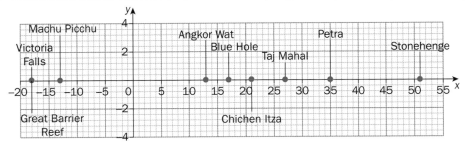

b 69 degrees   **c** 237 degrees

d Change in latitude: +35 degrees   **e** Latitude: +10 = +34 − 38 − 31 + 39 − 34 + 43 − 48 + 45

    Change in longitude: −114 degrees   Longitude: +166 = +86 + 106 − 78 − 115 + 16 + 108 + 113 − 70

15 a 10   **b** −9   **c** −16

Unit 4

Practice 1

1

Expression	Name	Degree
$-18x^3y$	Monomial	4
$-9x^2 + 2x - 21$	Trinomial (quadratic)	2
$-3y^5z^6 + 9yz^{11}$	Binomial	12
$11m^4n^4 - 7k^8$	Binomial	8
$\frac{5x-2}{3}$	Binomial	1
$12y^3 + 4y + 8$	Trinomial (cubic)	3
Individual response	Binomial	5
Individual response	Polynomial	1

2 a $-4x^3 + 8x + 7$ **b** $x^4 + 7x^3 - 2x^2 - x - 11$ **c** $-x^2 + x - 1$ **d** $2x^5 - 3x^4 - 4x^3$

    Degree 3   Degree 4   Degree 2   Degree 5

   e $-x^2 + 12$ **f** $x - 11$ **g** $-2x^3 + 3x^2 - 8x + 4$ **h** -8

    Degree 2   Degree 1   Degree 3   Degree 0

3 Individual responses

4 Across: Down:

3 Degree	1 Quintic
5 Linear	2 Quartic
9 Quadratic	4 Expression
10 Constant	6 Trinomial
12 Monomial	7 Cubic
13 Polynomia	8 Binomial
	11 Term

Practice 2

1 **a** $2x^2$ **b** $6x$ **c** 8 **d** $2xyz$

2 **a** $8x - 8y$ **b** $-7t^3 + 3r + 7$ **c** $-11t + 5k$ **d** $-21x^2y^2 + 17x^2y + xy^2$
   e $11xy - 4x - 9$ **f** $-2pq - 4q - 4$ **g** $-10ab + 16a - 8b$ **h** $-6x^2 + 3x - 21$
   i $-3k^2 - 6k + 17$ **j** $4a^2 - 7b^2 - 19a - 2b - 10$
   k $-7p^2q - 11pq^2 + 5p^2 + 6pq$ **l** 0

3 **a** The sum of each row is 15
   The sum of each column is 15
   The sum of each diagonal is 15
   b $15 = 5 \times 3$
   c $45 = 9 \times 5$. If the number in the middle is n then the numbers around it are $n - 4$, $n - 3$,
   $n - 2$, $n - 1$, $n + 1$, $n + 2$, $n + 3$, $n + 4$ and the sum of these numbers is $9n$. Setting $n = 5$ gives 45. A calendar is
   not a magic square.

4 **a**

2	9	4
7	5	3
6	1	8

6	7	2
1	5	9
8	3	4

   b Individual response

5 **a** Left: the rows, columns and diagonals all sum to $3x$
   Right: the rows, columns and diagonals all sum to $3x$
   b

−2	5	0
3	1	−1
2	−3	4

4	−11	10
7	1	−5
−8	13	−2

6 **a**

$-2x - y$	$4x - 2y$	$x + 3y$
$4x + 4y$	x	$-2x - 4y$
$x - 3y$	$-2x + 2y$	$4x + y$

   b

$3x + 2y$	$-7y$	$5y$
$-2x + 3y$	x	$4x - 3y$
$2x - 5y$	$2x + 7y$	$-x - 2y$

Practice 3

1 **a** $2x - 24$ **b** $45a - 30b$ **c** $6g - 6h$ **d** $20n + 17$ **e** $7w - 52$
   f $3x - 8$ **g** $6k - 34$ **h** $-5s + 18$ **i** 64 **j** $-33X + 28Y + 9$
   k $-11a + 4v - 7$ **l** $18x^3 + 2x^2 + 4x - 4$ **m** $-13i - 9j - 39$ **n** $-3x^2y^2 - 41x^2y - 45x + 16$

2 **a** Each of the rows, columns and diagonals sum to $6a - 9b$
   b Each of the rows, columns and diagonals sum to $-6a + 12b$
   c Each of the rows, columns and diagonals sum to $-27a + 9b$
   d Each of the rows, columns and diagonals sum to $6a^2 - 12a + 18$

e (a)

−6	−5	23
33	4	−25
−15	13	14

Rows/columns/diagonals sum to 12

(b)

3	−15	−6
−15	−6	3
−6	3	15

Rows/columns/diagonals sum to −18

(c)

−9	5	13
25	3	−19
−7	1	15

Rows/columns/diagonals sum to 9

(d)

−6	42	0
18	12	6
24	−18	30

Rows/columns/diagonals sum to 36

3 a, b, c Individual responses

4 a $\dfrac{4(n-3)+32}{4} - n = 5$ **b** $\dfrac{2(2(n+21)-2)+20}{4} - 25 = \dfrac{4n+100}{4} - 25 = n$ **c** $\dfrac{\frac{6(n+8)+2}{2}+2}{3} - n = \dfrac{3n+27}{3} - n = 9$

Practice 4

1 Note: choice of letter(s) is arbitrary

  a $2n+6$ **b** $3n+3$ **c** $39-3n$ **d** bg

  e $17n$ **f** $\dfrac{n}{2}$ **g** $-16n-24$ **h** $\dfrac{11+6mn}{2}$

2 Individual response

3 a n

    $2n$

    $2n+10$

    $10n+50$

    $10n+50+s$

    $10n+s$

    Provided you are 10 or older (and less than 100!) and have fewer than 10 siblings this gives three-digits, the first two of which is n, your age, and the final digit is s, the number of siblings you have

  b Let the volunteer's age be $10m+n$ where m, n are integers and $0 \le n \le 9$

    m

    $2m$

    $2m+3$

    $10m+15$

    $10m+15+n$

    $10m+15+n-15$

    $10m+n$

  c Let the birthday day be a, the birthday month be m and today b

    $20a$

    $20a+2b$

    $100a+10b$

    $100a+10b+m$

    $100a+m$

4 Individual Response

5 If the digits chosen are m and n two possible two–digit numbers are $10m+n$ and $10n+m$. The sum of these is $11(m+n)$ so dividing by $(m+n)$ gives 11 with certainty

6 a If the sum of the digits is divisible by 3 (respectively 9) then the number is divisible by 3 (respectively 9).

  b $100A+10B+C$

  c $100A+10B+C = 99A+A+9B+B+C = 99A+9B+A+B+C$

    This can be expressed as $9(11A+B)+A+B+C$ or $3(33A+3B)+A+B+C$

    so if the sum of the digits is divisible by 3 or 9, then so is this number

7 The list is given by a, b, $a+b$, $a+2b$, $2a+3b$, $3a+5b$, $5a+8b$, $8a+13b$, $13a+21b$, $21a+34b$

  The sum of these numbers is $55a+88b = (5a+8b) \times 11$

Practice 5

1 a $4b = 8 \Rightarrow b = 2$ **b** $2b + 7 = 5b + 4 \Rightarrow b = 1$ **c** $b + 3 = 2b + 1 \Rightarrow b = 2$ **d** $3b + 2 = b + 2 \Rightarrow b = 0$

2 a $x = 5$ **b** $p = -3$ **c** $z = -1$ **d** $s = -2$
   e $m = 5$ **f** $g = -3$ **g** $c = -4$ **h** $x = 20$
   i $x = 1$ **j** $x = -1$ **k** $x = 36$ **l** $w = 1$
   m $y = -20$ **n** $a = 9$ **o** $t = \dfrac{49}{8}$ **p** $d = 8$

3 a $m = -3$ **b** $m = -3$ **c** $m = -3$ **d** Individual response

Practice 6

1 a $n + (n+1) + (n+2) + (n+3) = 4n + 6$
$4n + 6 = 102 \Rightarrow n = 24$
So you chose 24, 25, 26, 27
   b n
$4n$
$4n + 3$
$6n + 3$
$6n - 6$
$n - 1$
You originally chose 41
   c $2n - (n + 3) = n - 3$
You originally chose 70, 71, 72, 73

2 a $n + (n+1) + (n+2) = 2(n+3) + 1$
$3n + 3 = 2n + 7 \Rightarrow n = 4$
The original numbers were 4, 5, 6
   b $n + (n + 6) = 2n + 6 = 5n \Rightarrow n = 2$
So she is referring to 2 and 8

3 $2n - 4 = 3n + 5 \Rightarrow n = -9$

4 Individual response

Practice 7

1 a $y = \dfrac{x}{2} + 5$ **b** $y = 5x - 6$ **c** $y = -\dfrac{x}{6} - 2$ **d** $y = 4 - 2x$
   e $y = -\dfrac{x}{2} + 7$ **f** $y = 3x + 2$ **g** $y = 3x - 15$ **h** $y = 12 - 4x$

2 a $y = \dfrac{x}{3} - 7$ **b** MATHEMATICS MAKES THE WORLD GO AROUND

Practice 8

1

2 a $n > -2$ **b** $n \leq 7$ **c** $n \geq 3$ **d** $n > 0$ **e** $n < -4$ **f** $n \leq -4$

3 a

b

c

d

Practice 9

1 a $x < 4$

b $t > -3$

c $m \leq 3$

d $g \geq -6$

e $y > -5$

f $p \leq 2$

g $w \geq 0$

h $w \leq 2$

i $v > -1$

j $w \leq -2$

k $y \leq -1$

l $d \leq -2$

2 Individual response

Unit Review

1 a Three terms, degree 3, quadratic   **b** Two terms, degree 1, linear
   c Four terms, degree 8, polynomial   **d** One term, degree 3, cubic

2 a $-5w^3y^4 + 16w^2y$ **b** $-3m^3 + 3m - 4$ **c** $-g^2h - 3gh^2$ **d** $-12f^2 - 11f + 8$

3 a $z = 4$ **b** $d = -4$ **c** $w = -2$ **d** $z = -12$

4 a $w = 2$ **b** $t = -\dfrac{1}{2}$ **c** $r = \dfrac{7}{5}$

5 a $v > -7$

     -8 -7 -6 -5 -4 -3 -2 -1 0 1 2 3 4 5 6 7 8

b $g \le -2$

     -8 -7 -6 -5 -4 -3 -2 -1 0 1 2 3 4 5 6 7 8

c $x \ge 6$

     -8 -7 -6 -5 -4 -3 -2 -1 0 1 2 3 4 5 6 7 8

d $x < 10$

     -12 -10 -8 -6 -4 -2 0 2 4 6 8 10 12

6 n
$$n + (n+1) = 2n + 1$$
$$2n + 1 + 7 = 2n + 8$$
$$(2n + 8) \div 2 = n + 4$$
$$n + 4 - n = 4$$

7 a $b = -5$ **b** $a = 6$

8 a $t > -3$ **b** $g \ge -2$ **c** $a > 8$

9 a $n + (n + 2) = 2(n + 1)$ **b** $n + (n + 3) = 2n + 3 = (n + 1) + (n + 2)$
   c $n + (n + 4) = 2n + 4 = 2(n + 2) = (n + 1) + (n + 3)$

10 16, 21

11 a $g = 8$ **b** $x = -2$

12 a Let the original number be $10a + b$ where $0 \le a, b \le 9$
     $10a + b - (10b + a) = 9a - 9b = 9(a - b)$

b Let there be $2m + 1$ consecutive integers $\{n, n + 1, ..., n + 2m\}$ so that the middle integer is $n + m$. Then the mean of these numbers is

$$\frac{n + (n + 1) + ... + (n + 2m)}{2m + 1} = \frac{(2m + 1)n + \frac{1}{2}(2m)(2m + 1)}{2m + 1} = n + m$$

     as required

13 12, 13, 14

14 Individual response

15 Half of the even number

16 Individual response

Unit 5

Practice 1

1 a Perimeter: 73.5 units   **b** Perimeter: 75 units   **c** Perimeter: 144 units   **d** Perimeter: 80 cm
     Area: 320 units^2 Area: 431.3 units^2 Area: 1562.4 units^2 Area: 492 cm^2

   e Perimeter: 126 units   **f** Perimeter: 90 cm   **g** Perimeter: 48 in   **h** Perimeter: 21 m
     Area: 1241.1 units^2 Area: 616.5 cm^2 Area: 168 in^2 Area: 24 m^2

2 Apothem: $4 \times (0.5 \times 4.4 \times 2.2)\,\text{cm}^2 = 19.36\,\text{cm}^2$
   Standard formula: $4.4^2\,\text{cm}^2 = 19.36\,\text{cm}^2$

3

Polygon	Side length (cm)	Apothem (cm)	Area (cm²)
Pentagon	6	4.1	61.9
Hexagon	10	8.7	259.8
Octagon	4.1	5	82.8
Decagon	20	30.8	3077.7
Hexagon	13.9	12	498.8
Pentagon	43.6	30	3269.4

4 a Regular octagon   **b** 6.0

5 a 340 km   **b** The borders of North Dakota, as with all territories, is a human defined landscape.

6 a 2340 cm²   **b** 4.5×10^{12}

7 a 20705.9 m²   **b** 114872.2 m²

Practice 2

1 a Circumference: 25.12 cm
Area: 50.24 cm²
b Circumference: 38.94 in
Area: 120.70 in²
c Circumference: 56.52 ft
Area: 254.34 ft²
d Circumference: 72.22 m
Area: 415.27 m²
e Circumference: 17.90 cm
Area: 25.50 cm²
f Circumference: 4.71 in
Area: 1.77 in²

2 a 107.1 m   **b** Individual response

3 a $2(84.39) + 2\pi(36.8) = 400.00$ **b** 7.04 m

Practice 3

1 a 72 cm²   **b** 96 ft²   **c** 240 cm²   **d** 907.2 cm²
e 1020 cm²   **f** 2352 mm²   **g** 352 cm²   **h** 386.50 m²

2

Prism	Side length (cm)	Apothem (cm)	Height of prism (cm)	Surface area (cm²)
Pentagonal	10	6.9	18	1245
Hexagonal	7	6.1	7	550.2
Heptagonal	12	12.5	21	2814
Octagonal	20	24.1	5	4656
Decagonal	16	24.6	14	6176

3 a 19.25 m²   **b** Individual response

4 a 152.28 m²   **b** 26.0%   **c** 8.88 m

Practice 4

1 a 120 ft³   **b** 200 ft³   **c** 2560 m³   **d** 206.3 m³   **e** 136.5 cm³   **f** 56 cm³

2

Length (cm)	Width (cm)	Height (cm)	Volume (cm³)
12	4	5	240
9	**20**	7	1260
2.4	4.9	6.5	78
4.1	7.6	**0.8**	23.4
18.1	12.5	6	**1357.5**
0.7	4.8	7.2	25.6

3 **a** 309 cm³ **b** 792 cm³ **c** 168 cm³

d 13 ft³ **e** 1512 ft³ **f** 192 ft³

4 **a** 512 cm³ **b** 11 cm **c** 10851.84 euros

5 **a** 39 750 000 m³ **b** 99 375

Practice 5

1 **a** **i** Rectangle, of area 80 cm² **b** **i** Rectangle, of area 104 mm² **c** **i** Pentagon, of area 240 ft²
   ii 960 cm³ **ii** 728 mm³ **ii** 5280 ft³
   d **i** Rectangle, of area 140 cm² **e** **i** Hexagon, of area 128.1 cm²
   ii 440 cm³ **ii** 1793.4 cm³

2

Shape	Base dimensions	Height of 3D figure	Volume
Rectangular Prism	Length = **0.3 mm** Width = 25 mm	35 mm	280 mm³
Triangular Prism	Base = 12 mm Height = 7 mm	**6.8 mm**	286 mm³
Pentagonal Prism	Side Length = 8.2 m Apothem = **2.2 m**	10.1 m	451.8 cm³
Rectangular Prism	Length = 21 m Width = **4.4 m**	14.2 m	1300 m³
Hexagonal Prism	Side Length = 5 mm Apothem = 3 mm	**2.5 mm**	112 mm³

3 **a** 3750 m³ **b** 3375 m³

4 **a** 210.5 mm³ **b** 7600.3

Unit Review

1 **a** Perimeter: 42 m **b** Perimeter: 260 ft **c** Perimeter: 40 in **d** Perimeter: 100.8 mm
   Area: 63 m² Area: 5460 ft² Area: 320 in² Area: 756 mm²
   e Perimeter: 79 cm **f** Perimeter: 144 cm **g** Perimeter: 38π cm **h** Perimeter: 8.4 m
   Area: 477.95 cm² Area: 1584 cm² Area: 361π cm² Area: 17.64π m²

2 **a** Surface Area: 212.4 cm² **b** Surface Area: 63 m²
   Volume: 194.4 cm³ Volume: 15 m³
   c Surface Area: 588 mm² **d** Surface Area: 727.2 m²
   Volume: 1008 mm³ Volume: 234 m³

3 **a** Perimeter: 29.2 cm **b** Perimeter: 28 in
   Area: 42.5 cm² Area: 40 in²

4

Polygon	Side length (cm)	Apothem (cm)	Area (cm²)
Pentagon	9	4.1	92.3
Hexagon	4.8	10	144
Octagon	24	7.0	672.5
Decagon	30	8	1200

5 1038 cm²

6 **a** Surface Area: 300 cm² **b** Surface Area: 2028.8 ft²
   Volume: 304 cm³ Volume: 6828.8 ft³
   c Surface Area: 351.0 mm² **d** Surface Area: 306 m²
   Volume: 390 mm³ Volume: 214.5 m³

7 **a** 1375 m² **b** 55 m **c** 18 m

8 a $8659.0\,\text{m}^2$ **b** $329.9\,\text{m}$ **c** $53.0\,\text{m}$

9 a $855.3\,\text{m}^2$ **b** Surface Area: $30.9\,\text{m}^2$
                             Volume: $284.1\,\text{m}^3$

10 a $2\,144\,767.5\,\text{feet}^3$
    b Individual response (e.g. flooring/ceiling on each floor takes up volume, need for corridors etc.)
    c 10.45% (accurate to 2 decimal places)

11 a Triangular prism with base width of $69\,\text{km}$, base height of $11\,\text{km}$, and length of $2\,500\,\text{km}$
    b $948\,750\,\text{km}^3$

12 a $1\,800\,000\,\text{m}^3/\text{h}$ **b** $3.25\,\text{m}$

13 a $496\,800\,000\,\text{km}^2$ **b** 1.42%   **c** $4.968 \times 10^{10}\,\text{km}^3$ **d** 1.42%

14 a $175\,000\,\text{km}^3$ **b** Individual response

15 2

Unit 6

Conversions used throughout this chapter: 1 foot = 12 inches, 3 feet = 1 yard, 5280 feet = 1 mile, 2.54 cm = 1 inch, 1 mile = 1.61 km

Practice 1

1 a 1.9   **b** 23.6   **c** 3520   **d** 365.8   **e** 0.1   **f** 16397.5

2 a 180   **b** 46.7   **c** 60   **d** 50   **e** 6.7   **f** 6

3 a 856   **b** 10.2 seconds

4 a $6169\,\text{kg}$ (to the nearest kg)   **b** $16\,131\,\text{kg}$ (to the nearest kg)   **c** 20 088 liters (to the nearest liter)
    d 4917 liters   **e** 15 171 liters

Practice 2

1 a 1760.4 Dominican pesos   **b** 2330.10 Australian dollars   **c** 72.6 US dollars
    d 45.84 euros   **e** 68.42 euros   **f** 65.79 Malaysian ringgit

2 a Hong Kong   **b** New Zealand

3 a 2 sheep = 25 loaves of bread   **b** 75 loaves of bread
    c 3.2 sheep   **d** Individual response (e.g. not feasible to trade a fraction of a sheep)

4 a 35 knives = 2 beaver skins   **b** 175 knives
    c 2.9 beaver skins   **d** 4 beaver skins

Practice 3

1 a $3:1$ **b** $0.2:1$ **c** Unit rate   **d** $3:1$ **e** $15:1$ **f** $0.5:1$
    g $2:1$ **h** $8:1$ **i** $6:1$ **j** $0.6:1$ **k** $19:1$ **l** $0.2:1$

2 a 1.5 meters per second   **b** −4 degrees per hour
    c 2 euros per bag   **d** 36 Thai baht for 1 pair of sunglasses
    e 0.125 movie tickets per British pound   **f** 12.5 kilometers per hour
    g 5 cm of snow per hour   **h** 6 staff members for 1 train

3 Kiara

4 22.73 pesos

5 a $5:1$ **b** $1.5:1$ **c** $0.6:1$ **d** $0.125:1$ **e** $5:1$ **f** $3.5:1$

6 a Left   **b** Right   **c** Right   **d** Left

7 a First option   **b** First option   **c** First option   **d** First option   **e** First option

8 a 3065.13 US dollars   **b** 2962.96 US dollars   **c** Individual response

Practice 4

1 0.8 hours

2 16 hours

3 280 litres

4 **a** 69.8 years   **b** Individual response

5 20 minutes

6 **a** 26.25 hours   **b** 14 people

7 4 people × 3 meals per day × 5 days = 60 meals
   5 people × 3 meals per day × 3 days = 45 meals
   So, yes, there will be plenty of food for the five of you.

Practice 5

1 **a** Constant rate of change   **b** Not a constant rate of change
   c Not a constant rate of change   **d** Constant rate of change

2 **a** Constant rate of change   **b** Not a constant rate of change
   c Constant rate of change   **d** Not a constant rate of change

3 **a** Constant rate of change   **b** Not a constant rate of change
   c Constant rate of change   **d** Not a constant rate of change
   e Constant rate of change   **f** Not a constant rate of change

4 Individual response

5 **a** Individual response   **b** Yes   **c** Yes, $t = 0.155p$

6 **a** 4184 J = 1 kcal   **b** Yes   **c** Yes $k = \dfrac{1}{4184} J$

7 **a**

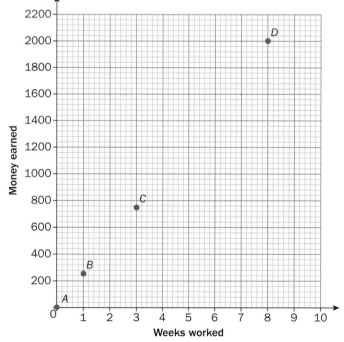

   b Yes, 250 units of money earned per week of work   **c** Yes   **d** $y = 250x$

Unit Review

1 **a** 7.62 meters   **b** 42.2 km   **c** 32.8 yards   **d** 0.1 litres   **e** 180 mL   **f** 33.3 tablespoons

2 **a** 20 miles per hour   **b** 15 kilometres per hour   **c** $\dfrac{2}{3}$ inch per day   **d** 0.375 miles per minute

3 a Constant rate of change   **b** Not a constant rate of change
  c Constant rate of change   **d** Constant rate of change

4 13.6 km /liter

5 14.1 laps

6 a 540 Uruguayan pesos   **b** 11.39 Australian dollars   **c** 89 550 Lebanese pounds
  d 5000 Rwandan francs   **e** 4.35 Belize dollars   **f** 2380.95 Barbadian dollars

7 a 6500 Uruguayan pesos   **b** 12 British pounds
  c 150 UAE dirhams   **d** 50 Yemeni rials

8 23.1 grains

9 a 6 hours   **b** 3 hours 20 minutes

10 a

Time (Weeks)	Money earnt (Brazilian reals)
1	1600
2	3200
3	4800
4	6400
5	8000
6	9600
7	11200
8	12800

  b Money earnt (in Brazilian reals) plotted against time (in weeks). Straight line passing through origin with gradient 1600
  c Let the money earnt be represented by m and time in weeks be represented by w. Then, $m = 1600w$
  d After 1 week, Joseph earnt 1600 Brazilian reals

11 a 675 pounds of force   **b** 12888.9 newtons

12 a

Day	Tatal spent ($)
1	100
2	200
3	300
4	400
5	500
6	600
7	700
8	800
9	900
10	1000
11	1100
12	1200
13	1300
14	1400
15	1500
16	1600
17	1700
18	1800
19	1900
20	2000

  b Dollars spent plotted against days on vacation. Straight line through origin, with gradient 100, from (0,0) to (20,2000).
  c Let money spent be represented by m and days sept on vacation be represented by d. Then, $m = 100d$
  d This point represents how much money in total Rebecca has spent after 20 days on her vacation.

Unit 7

Practice 1

1 a

Stem	Leaf
10	0 1 2 4
11	1 2 2 3 7 7 8
12	0 3 7

Key: 10 | 0 = 100

b

Stem	Leaf
5	1 2
6	
7	2 6
8	1 4 8
9	0

Key: 5 | 1 = 5.1

c

Stem	Leaf
17	1
18	0 2
19	4 5
20	3 5 6
21	5

Key: 17 | 1 = 17.1

d

Stem	Leaf
111	1 3 7 9
112	0 5 6 6 8
123	5

Key: 111 | 1 = 1111

2 a 2   **b** 13   **c** 21   **d** Individual response

3 a

	Stem	Leaf
9	0	
0 0 0 1 2 2 3 3	1	
	2	7 9 9
	3	1 2 2 3 3 4

Key: 2 | 7 = 27 kg

b

	Stem	Leaf
	0	8 9
2 5	1	0 0 2 4 5 6
1 5 6 8	2	
0 0	3	

Key: 1 | 10 = 10 students

c

	Stem	Leaf
	11	9
	12	5 8
	13	1 2 3 8
8	14	8
0 2 4 9	15	0
1 1 6	16	
3	17	

Key: 11 | 9 = 119 cm

4 a (Units in kcal per capita)
  Low-income: Highest: 2520, Lowest: 2020
  High-income: Highest: 3640, Lowest: 3000
  b No; the lowest crop supply in high-income countries is greater than the highest crop supply in low-income countries
  c High-income: 3490, Low-income: 2150
  d Individual response (all answers should give evidence that there is not equal access to food supplies)

5 a Individual response (e.g. low-income countries on average have more new HIV cases than high-income countries)
  b Individual response
  c Individual response (e.g. high-income countries are more likely to have the capacity to provide a better quality of education and treatment)
  d Individual response
  e Individual response

Practice 2

1 a Mode: None
  Median: 20
  b Mode: 4.5
  Median: 4.5
  c Mode: 2, 4 and 7
  Median: 4.5
  d Mode: 80
  Median: 72.5
  e Mode: 21 100
  Median: 21 600
  f Mode: 6
  Median: 6
  g Mode: 9.8 and 10.1
  Median: 9.8
  h Mode: 72
  Median: 73

2 a Two modes: 97 and 99
  Median: 96.5%
  b Individual response
  c Individual response (most probably Haiti)
  d Median: 97%
  Mode: 97% and 99%

e Individual response (e.g. increase in median, no effect on mode)
f Individual response

3 a Mean: 98.46% to nearest hundredth   **b** Individual response   **c** Individual response
Median: 100%

4 a Median: 17   **b** Individual response (e.g. 10 or 24)   **c** Individual response (17)   **d** Yes, e.g 20
Two modes: 17 and 20

5 Individual response

Practice 3

1 a Mean: 33.6   **b** Mean: 6.9   **c** Mean: 114.3   **d** Mean: 9.9
Median: 35.5   Median: 6.9   Median: 116.5   Median: 9
Mode: 24   Mode: 6.9 and 7.1   Mode: None   Mode: 9

2 a Mean: 2.8   **b** Individual response   **c** Individual response   **d** Individual response
Median: 3
Mode: 2 and 4

3 a Mean: 3.3   **b** Individual response   **c** Individual response
Median: 3
Mode: 3
   d Individual response (predominantly the mean)   **e** Median
   f Individual response (e.g. longer waiting times when ill)

4 Individual response

5 Individual response

6 Individual response

Practice 4

1 a Range: 17
Five-number summary: 25, 29.5, 35.5, 40 ,42
Interquartile range: 10.5
  b Range: 9
Five-number summary: 11.4, 13.4, 16, 18.0, 20.4
Interquartile range: 4.6
  c Range: 60
Five-number summary: 300, 310, 340, 360, 360
Interquartile range: 50
  d Range: 4.5
Five-number summary: 1.1, 2.4, 3.2, 3.9, 5.6
Interquartile range: 1.5
  e Range: 270
Five-number summary: 1270, 1275, 1430, 1475, 1540
Interquartile range: 200
  f Range: 24
Five-number summary: 430, 432.5, 441.5, 449, 454
Interquartile range: 16.5
  g Range: 5
Five-number summary: 3, 4, 5, 7.5, 8
Interquartile range: 3.5

2 a 'High-income countries' five-number summary: 34%, 35%, 38.5%, 45%, 58%
'Low-income countries' five-number summary: 44%, 47%, 51%, 55.5%, 59%
  b Individual response (low-income countries appear to have better availability of arable land)
  c Individual response

3 a Range: 973   **b** Yes, Singapore and Italy; these are far above the rest of the data set
Lower Quartile: 10
Upper Quartile: 27
Interquartile range: 17

c Range: 34
Lower Quartile: 10
Upper Quartile: 13
Interquartile range: 3

d Range (range decreased by 939 whereas interquartile range only decreased by 14)

4 Individual response

5 Individual response

6 Individual response

Practice 5

1 a Range: 19
Lower quartile: 14
Median: 19
Upper quartile: 23
Interquartile range: 9

b Range: 39.6
Lower quartile: 63
Median: 65
Upper quartile: 70
Interquartile range: 5

c Range: 14
Lower quartile: 104
Median: 108
Upper quartile: 110
Interquartile range: 6

2 a 0.2 km   **b** 5 km   **c** 25%   **d** Individual response   **e** Individual response

3 a
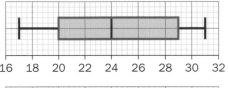

b
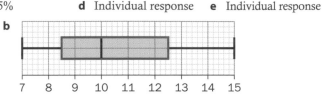

c
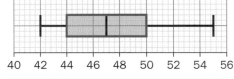

d
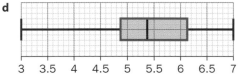

e
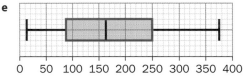

f
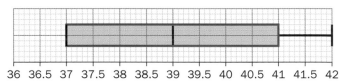

4 a
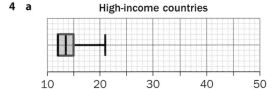

b Birth rates in low-income countries are significantly above the birth rates in high-income countries

c Individual response

5 a
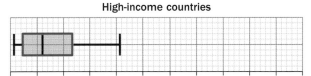
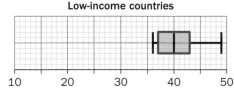
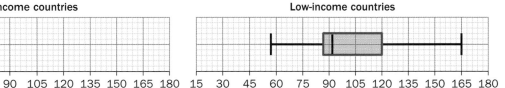

b In general, there are a greater number of threatened species of birds in low-income countries than in high-income countries

c Individual response (e.g. Not necessarily, but it is dependent on the reason that the species is threatened. For example, if the low-income country must rely on an industry that destroys the birds' habitats or kills them indirectly via pesticides then yes. Conversely, external influences such as non-native predators or toxic contaminants / plastics flowing inshore could cause threat to the species.)

d Individual responses

6 a, b, c Individual response

Unit Review

1 a

Stem	Leaf
6	1 1 4 4 5
7	0 2 3 9 9
8	0 3

Key: 6 | 1 = 61

b

Stem	Leaf
3	0 4 9
4	1 6 7 9
5	1 5 5
6	8

Key: 3 | 0 = 3.0

c

Stem	Leaf
12	00 00 10 60 70 80
13	50 50 60 90
14	00 10
15	60

Key: 12 | 00 = 1200

2 a Mean: 70.9
Median: 71
Two modes: 61 and 79

b Mean: 4.7
Median: 4.7
Mode: 5.5

c Mean: 1326.2
Median: 1350
Two modes: 1200 and 1350

3 a Range: 22; Lower Quartile: 64; Upper Quartile: 79; Interquartile Range: 15
b Range: 3.8; Lower Quartile: 3.9; Upper Quartile: 5.5; Interquartile Range: 1.6
c Range: 360; Lower Quartile: 1235; Upper Quartile: 1395; Interquartile Range: 160

4 a Highest: 25.5%
Lowest: 18.0%

b 10

c Mean: 20.8
Median: 20.5
Mode: 18.6

5 a

High-income countries

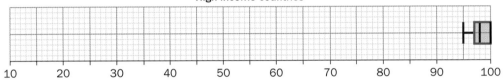

Low-income countries

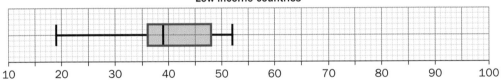

b Individual response

6 a

Stem	Leaf
18	0 0 0 5 5 5
19	0 0 0 2 2
20	0 0 0
21	6 6
22	
23	
24	0 0

Key: 18 | 0 = 180 days of school per year

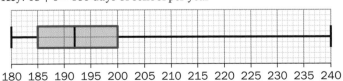

b Individual response **c** Individual response **d** Individual response

7 a Low-income countries: Maldives and Bangladesh
High-income countries: Monaco

b Low-income countries with outliers: Mean: 721.2
    Median: 587
    Mode: None
Low-income countries without outliers: Mean: 549.6
    Median: 526
    Mode: None
High-income countries with outliers: Mean: 4202.9
    Median: 1365
    Mode: None
High-income countries without outliers: Mean: 2322
    Median: 1336
    Mode: None

c Individual response **d** Individual response **e** Individual response **f** Individual response

8 a 2200 AUD **b** 9000 AUD **c** Individual response
   Individual response
d The new box-and-whisker plot would have the same lower value, but the much higher new upper value would impact the median and the two quartiles.
e Individual response

9 a Low-income countries: Smallest value = 16%, Q1 = 19%, Median = 23%, Q3 = 27.5%, Largest value = 29%
   High-income countries: 89%, 90.5%, 92%, 96.5%, 97%

b

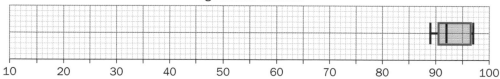

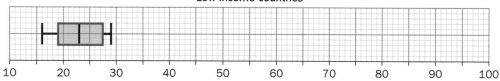

c Individual response (e.g. High-income countries have greater access to improved sanitation in urban areas than low-income countries)
d Low-income countries: Mean: 22.9%   High-income countries: Mean: 93.3%
    Median: 23%   Median: 92%
    Mode: 20%   Mode: 92% and 96% and 97%
e Individual response
f Individual response (e.g. Yes the mean/median/mode of the high-income countries is a significantly higher percentage than the mean/median/mode for the low-income countries, suggesting an unfairness in access to improved sanitation)
g Individual response   **h** Individual response

10 Individual response

11 a 49%   **b** 8.5%

12 a 37.4   **b** Individual response   **c** 30.3
d Individual response (e.g. an extreme value artificially shifts the mean towards the direction in which the value is extreme)
e Individual response
f

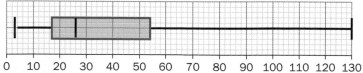

Qatar appears to be an outlier.

Index